PAINTING AND DECORATING PRICE BOOK
19th Edition 2012

CONSTRUCTION

BCIS is the Building Cost Information Service of

BCIS
50 years celebrating excellence

CONSTRUCTION · MAINTENANCE COSTS & REPAIRS · REBUILDING COSTS · INTELLIGENCE · 2012

HOW LONG? HOW MUCH?
THE FASTEST, MOST UP-TO-DATE ANSWERS ARE AVAILABLE NOW

Cost information underpins every aspect of the built environment, from construction and rebuilding to maintenance and operation publications.

BCIS, the RICS' Building Cost Information Service, is the leading provider of cost information to the construction industry and anyone else who needs comprehensive, accurate and independent data.

For the past 50 years, BCIS has been collecting, collating, analysing, modelling and interpreting cost information. Today, BCIS make that information easily accessible through online applications, data licensing and publications.

For more information call **+44 (0)870 333 1600** email **contact@bcis.co.uk** or visit **www.bcis.co.uk**

BCIS is the Building Cost Information Service of  RICS

the mark of property professionalism worldwide

PREFACE
FOREWORD TO THE NINETEENTH EDITION

This new edition of the BCIS Painting and Decorating Price Book has the revised style and layout. We hope you will find it clearer, quicker and easier to use, especially with the enhanced index. Prices and specifications have been thoroughly revised and are current as at December 2011.

The aims of this edition remain the same as its predecessors; that is, to provide a quick source of reference for painting and decorating prices and information appertaining to the trade.

This Price Book contains guide prices for redecoration work in existing properties as well as for decoration work in new buildings. It includes internal and external work and a full range of surface treatments. The pricing information is presented in two sections; firstly, in the form of prices "per operation" or "per coat", enabling the user to build up an infinite number of unit rates, by combining these prices to meet a particular job specification and secondly, in the form of composite, or compound, prices in which total rates are given for a range of the most common painting and decorating specifications for rapid estimate calculation. This method provides the user with a flexible and easy-to-use Database.

For every price, the detailed labour content and material usage is shown. A summation of hours involved in the various items included in the specification items will give the total labour resources required for the job to enable work planning, scheduling and sequencing to be carried out for large or small contracts. Similarly, a review of the material usage and prices will enable quantities to be easily calculated and optimum supply packs and discounts obtained.

Thus, the BCIS Painting and Decorating Price Book is for the small firm, the large firm, the new firm or the established concern. Contractors, surveyors, estimators, interior designers, in fact, everyone involved in the trade will find it a valuable ready reference.

N P Barnett BSc (Hons)
Information Services & Cost Resource Manager
BCIS

COPYRIGHT
BCIS PAINTING & DECORATING PRICE BOOK 2012

COPYRIGHT

(c) copyright 1985, 1988, 1990, 1995, 1998, 1999, 2000, 2001, 2002, 2003, 2004, 2005, 2006, 2007, 2008, 2009, 2010, 2011, 2012

BCIS

All rights reserved. No part of this publication may be reproduced, or stored on computer, or used as a Database for any computer system application or computer estimating system, or stored in a retrieval system or transmitted in any form or by any means, electronic, mechanical, photocopying, recording or otherwise without the prior written permission from the copyright owner.

Any infringement of the copyright in this publication will be treated most seriously and it is the Publishers' intention to take immediate legal action against any company, firm or individual guilty of infringement.

Although the greatest care has been taken in the preparation and compilation of the guide prices used in this publication the Authors, Publishers and Copyright owners can accept no responsibility for any errors or omissions or for any claims arising from the use of this book. It is the Publishers' policy to continue to improve and develop the BCIS Databases and we invite your suggestions for improvements to the book.

ISBN: 978 1 907196 24 9

BCIS
Suite 1, Ground Floor
376 Ringwood Road
Poole, Dorset
BH12 3LT

Telephone:	+44 (0)870 333 1600
Fax:	+44 (0)20 7695 1501
Web Site:	www.bcis.co.uk
E-Mail	contact@bcis.co.uk

February 2012

ACKNOWLEDGEMENTS
BCIS PAINTING & DECORATING PRICE BOOKS 2012

19TH EDITION

BCIS would like to thank our members of staff and all those who assisted in the preparation and production of the 19th Edition of the BCIS Painting & Decorating Price Book 2012. We also wish to acknowledge the invaluable assistance given by the following individuals, organisations, manufacturers, suppliers, contractors and associations, including those who have given their kind permission for the reproduction and publication of copyright material.

Technical Editor:
NP Barnett BSc (Hons)

Editorial staff and contributors to original material:
C A Rowe MRICS
A J Hobbs DMS,FRICS,FBIM
C Fleming MCIOB, M.Inst.C.E.S, MIAS
S G Price FRICS
K G Bray FFS, FFB
M J Snarr
Mrs M Damon–De Waele

Editorial Assistants:
Mrs C Barnett
Mrs R Read

Suppliers and Services:
Buttle PLC
Cornish Lime Company
Decorating Direct
ICI Paints AkzoNobel
Lawson HIS
Mackay
Muraspec Ltd
Screwfix
Tradec
Travis Perkins
Wood Care Direct

BCIS 50 years celebrating excellence

CONSTRUCTION

LESS DESK TIME MORE FREE TIME

THE REVOLUTION OF THE PRICE BOOK IS HERE
BCIS ONLINE RATES DATABASE

As a purchaser of the 2012 price book, we would like to offer you a free two week trial of our Online Rates Database. It's the price book but online, with lots of additional features to help you to:

- locate prices quickly using faster navigation
- adjust your data to suit a time frame of your choice and choose location factors to make your costs more accurate.

Plus:

- everything you need is in one place – you have a full library at your disposal
- with the full service you can download data into an Excel spreadsheet and manipulate and store your data electronically*.

*During the two week trial data downloads are not available

Offering immediate online access to independent BCIS resource rates data, quantity surveyors and others in the construction industry can have all the information needed to compile and check estimates on their desktops. You won't need to worry about being able to lay your hands on the office copy of the latest price books, all of the information is now easily accessible online.

For a **FREE TRIAL**
of BCIS online rates database, register at **www.bcis.co.uk/ordbdemo**

- Accuracy
- Futureproof
- Value for money
- Saves time

- Flexible
- Customise
- Portable
- Comprehensive

BCIS is the Building Cost Information Service of  the mark of property professionalism worldwide

Contents

Category	Section	Page
PREFACE		Beginning
COPYRIGHT		Beginning
ACKNOWLEDGEMENTS		Beginning
INDEX		i
INTRODUCTION		1
GENERAL INFORMATION		9
BASIC RATES		17
NEW WORK - INTERNALLY	VA-VG	29
NEW WORK - EXTERNALLY	VH-VM	87
WALL COVERINGS	VN	117
REDECORATION - INTERNALLY	VO-VP	123
REDECORATION - EXTERNALLY	VQ	141
COMPOSITE PRICES	VZ	149
USEFUL ADDRESSES		169

RICSBOOKS.COM

Leading supplier of books and contracts for the surveying, construction and property profession.

Shop online at www.ricsbooks.com

- Free delivery on orders over £75
- Exclusive web offers & discounts
- Standard & next day delivery
- Open when you want 24/7

INDEX

INDEX

(Bold italic numbers refer to section numbers, plain numbers are page numbers)

-A-

Adjustments for roller application
    painting and decorating, p-19
Alkali resisting primer externally
    paint externally, p-88
Alkyd based paints internally
    eggshell finish, p-38
    eggshell finish, sprayed, p-53
    gloss finish, p-40
    gloss finish sprayed, p-54
    undercoat, p-37
    undercoat sprayed, p-52
Anti-bacterial paint
    internally, water based epoxy, p-44
Artex finish internally
    paint internally, p-34
Average quality wallpaper
    decorative wall coverings, p-119

-B-

Breather paint on woodwork externally
    composite rates for painting, p-165
Brush allowance, p-20
    General Information, p-11
Burning off paint externally
    previously alkyd painted woodwork, p-145
Burning off paintwork
    walls and ceilings, p-130
    woodwork, p-135

-C-

Chlorinated rubber paint
    externally, line marking, p-113
Clear finish on woodwork externally
    composite rates for painting, p-164
Clear finish on woodwork internally
    composite rates for painting, p-159
Clear finishes
    raw linseed oil, p-108
Composite rates for painting
    anti-bacterial painting system, p-153
    breather paint on woodwork externally, p-165
    clear finish on woodwork externally, p-164
    clear finish on woodwork internally, p-159
    eggshell paint finish internally, p-151
    emulsion paint internally, p-150
    emulsion paint on walls externally, p-161
    emulsion paint, sprayed externally, p-162
    externally, p-161
    floor painting system, p-160
    gloss paint finish internally, p-152
    hardwood finish externally, p-165
    internally, p-150
    intumescent flame retardant paint system, p-157
    intumescent painting system, eggshell timber, p-159
    intumescent painting system, timber, p-158
    intumescent painting system, varnish timber, p-160
    intumescent wallcoat system, p-154, 155
    line marking, p-166
    metalwork internally, p-156
    paint on masonry, externally, p-161
    paperhanging and decoration, p-166
    primers, internally, p-151
    railings, p-163
    Sandtex Matt - walls externally, p-162
    signwriting, p-163
    Snowcem paint on walls externally, p-161
    wood stain externally, p-165
    woodwork externally, p-163
    woodwork internally, p-158
Cotton backed vinyl wall coverings
    wall coverings, p-120
Cuprinol
    on woodwork, p-109
Cutting in to line
    painting internally, p-41

-D-

Decorative wall coverings
    average quality wallpaper, p-119
    heavy embossed wallpaper, p-118
    paper, blown nylon surfaced, p-119
    ready pasted vinyl surfaced wallpaper, p-119
    vinyl surfaced wallpaper, p-119
    wallpaper, p-118, 119
    wood chip paper, p-118

-E-

Eggshell finish
    composite rates for painting, p-151
    internally, p-38
    sprayed internally, p-53
Emulsion paint
    externally, p-88
    internally, p-30
    sprayed externally, p-91
    sprayed internally, p-48
Emulsion paint internally
    composite rates for painting, p-150
Emulsion paint on walls externally
    composite rates for painting, p-161
Emulsion paint, sprayed externally
    composite rates for painting, p-162
External redecoration
    fungicidal treatment, p-142

INDEX

generally, p-142
previously alkyd painted metalwork, p-143
previously alkyd painted woodwork, p-144
previously emulsion painted surfaces, p-142
Externally
composite rates for painting, p-161

-F-

Fibre cement
primer, externally, p-112
Floor paint
semi-gloss, painting internally, p-59
Floor paint, Thermoplastic
painting internally, p-59
painting internally, non-slip, p-59
Fungicidal treatment
redecoration externally, p-142

-G-

General information
brush allowance, p-11
health and safety, p-13
paint coverage, p-9
preparation for painting, p-11
Glass fibre
wall coverings, p-120
Glaze to multicolour surfaces
paint internally, p-39
Gloss finish internally
acrylated rubber on wood, p-79
alkyd based paints, p-40
sprayed alkyd based paint, p-54
Gloss paint finish
composite rates for painting, internally, p-152
on fibrous cement, externally, p-112
on metal, externally, p-99
on metal, internally, p-67
on wood, externally, p-106
on wood, internally, p-77
Green wood preserver on woodwork
preservatives, p-109

-H-

Hammerite
paint externally, p-101
paint internally, p-71
Hammerite smooth
paint internally, p-71
Hardwood finish externally
composite rates for painting, p-165
Health and safety
General Information, p-13

-I-

Internally
composite rates for painting, p-150
Introduction
labour costs, p-2
materials, p-5
overheads and profit, p-1
painting and decorating, p-1
regional variations, p-2
wage award, p-4
Intumescent paint
externally to metal, coloured, p-103
internally to metal, coloured, p-73
internally to walls, coloured, p-46
internally to wood, clear, p-83
internally to wood, coloured, p-80

-L-

Labour costs
Introduction, p-2
Labour rates
painting and decorating, p-19
Line marking paint
chlorinated rubber, p-113
Lining paper
wallpaper, p-118

-M-

Masonry paint on walls
externally, p-89
Masonry paint, sprayed on walls
externally, p-91
Masonry sealer
walls externally, p-88
Materials
Introduction, p-5
Metalwork
paint stripping, p-133
painting internally, p-68
primers internally, p-59
redecoration, p-132
Metalwork externally
primers, p-91
Metalwork internally
composite rates for painting, p-156
Multicolour finish - sprayed
paint internally, p-57, 58

-O-

Old walls and ceilings - preparation
repainting and redecoration work, p-124
Opaque woodstain on woodwork
preservatives, p-111

Overheads and Profit
    Introduction, p-1

-P-

Paint coverage
    General Information, p-9
Paint externally
    alkali resisting primer, p-88
    emulsion paint on walls, p-88
    emulsion sprayed on walls, p-91
    fibre primer, p-112
    gloss finish on fibrous cement, p-112
    gloss finish on metal, p-99
    gloss finish on wood, p-106
    Hammerite, p-101
    masonry paint - composite rates, p-161
    masonry paint on walls, p-89
    masonry paint, sprayed on walls, p-91
    primer acrylic on wood, p-105
    primer alkali on fibre, p-112
    primer alkali resisting on walls, p-88
    primer aluminium sealer on wood, p-104
    primer etching on metal, p-91
    primer red oxide on metal, p-95
    primer zinc chromate on metal, p-93
    primer zinc phosphate on metal, p-92
    primer zinc rich on metal, p-94
    Sandtex Matt on walls, p-90
    undercoat on fibrous cement, p-112
    undercoat on metal, p-98
    undercoat on wood, p-105
    varnish on wood, p-107
    walls externally, p-88
    woodwork externally, p-104
Paint internally
    acrylated rubber on walls, p-41
    acrylated rubber paint (undercoat), p-55
    acrylated rubber paint , sprayed, p-56
    acrylated rubber paint on wood, p-43
    artex finish internally, p-34
    cutting in to line, p-41
    emulsion on walls and ceilings, p-30
    emulsion paint, sprayed, p-48
    etching primer, p-92
    glaze to multicolour surfaces, p-39
    gloss on metal, p-67
    gloss on wood, p-77
    Hammerite, p-71
    Hammerite smooth, p-71
    metalwork, p-68
    multicolour finish - sprayed, p-57, 58
    primer acrylic on wood, p-75
    primer aluminium sealer on wood, p-74
    primer etching on metal, p-59

    primer on walls and ceilings, p-36
    primer red oxide on metal, p-63
    primer zinc phosphate on metal, p-60
    primer zinc rich on metal, p-62
    primer/sealer bonding coat - sprayed, p-57
    primers internally, p-35
    primers, woodwork internally, p-73
    undercoat on metal, p-66
    undercoat on wood, p-76
    walls and ceilings, p-30
    woodwork clear finish, p-81
Paint stripping
    metalwork, p-133
    walls and ceilings, p-131
    woodwork, p-136
Painting and Decorating
    adjustments for roller applied paints, p-19
    Introduction, p-1
    labour rates, p-19
Paper, blown nylon surfaced
    decorative wall coverings, p-119
Paperhanging and decoration
    composite rates for painting, p-166
Preparation for painting
    General Information, p-11
Preservatives on woodwork
    green wood preserver, p-109
    opaque woodstain, p-111
    red cedar wood preserver, p-110
    water repellent clear wood preserver, p-109
Previously alkyd painted metalwork
    external redecoration, p-143
Previously alkyd painted surfaces
    redecoration, p-129
Previously alkyd painted woodwork
    burning off paint externally, p-145
    external redecoration, p-144
Previously emulsion painted surfaces
    external redecoration, p-142
    internal redecoration, p-127
    redecoration, p-128
Previously size distempered surfaces
    redecoration, p-124
Previously stained woodwork surfaces
    redecoration, p-136
Primer acrylic on wood
    externally, p-105
    internally, p-75
Primer alkali resisting on walls
    externally, p-88
Primer aluminium sealer on wood
    externally, p-104
    internally, p-74

INDEX

Primer etching on metal
    externally, p-91
    internally, p-59
Primer on walls and ceilings
    internally, p-36
Primer red oxide on metal
    externally, p-95
    internally, p-63
Primer zinc chromate on metal
    externally, p-93
Primer zinc phosphate on metal
    externally, p-92
    internally, p-60
Primer zinc rich on metal
    externally, p-94
    internally, p-62
Primer/sealer bonding coat - sprayed
    internally, p-57
Primers
    internally, p-35
    on metalwork externally, p-91
    on metalwork internally, p-59
    on woodwork internally, p-73
Primers internally
    composite rates for painting, p-151

-R-

Raw linseed oil on wood
    clear finishes, p-108
Ready pasted vinyl surfaced wallpaper
    decorative wall coverings, p-119
Red cedar wood preserver on woodwork
    preservatives, p-110
Redecoration
    externally, p-142
    metalwork, p-132
    previously alkyd painted surfaces, p-129
    previously emulsion painted surfaces, p-127, 128
    previously size distempered surfaces, p-124
    previously stained woodwork surfaces, p-136
    water painted, lime washed surfaces, p-125, 126
    woodwork, p-134
Regional variations
    Introduction, p-2
Repainting and redecoration work
    old walls and ceilings - preparation, p-124

-S-

Sadolin
    'Classic' externally, p-111
    on woodwork, p-111
    'Extra' externally, p-111
    'Superdec' externally, p-111
Sandtex Matt to walls
    externally, p-90

Sandtex paint externally
    composite rates for painting, p-162
Signwriting, p-106
Sizing surfaces for
    wallpaper, p-118
Snowcem paint
    composite rates for painting, externally, p-161
    walls externally, p-89
Sprayed paint finishes
    walls and ceilings, internally, p-48
    walls externally, p-91
Staining
    wood internally, p-81
Stripping off
    wallpaper, p-137
Suede effect wall coverings
    wall coverings, p-120

-T-

Textile wall coverings
    wall coverings, p-119

-U-

Undercoat externally
    on fibrous cement, p-112
    on metal, p-98
    on wood, p-105
Undercoat internally
    acrylated rubber on wood, p-78
    alkyd based paints, p-37
    on metal, p-66
    on wood, p-76
Undercoat sprayed internally
    alkyd based paints, p-52

-V-

Varnish on wood
    paint externally, p-107
Vinyl surfaced
    wall coverings, p-120
Vinyl surfaced wallpaper
    decorative wall coverings, p-119

-W-

Wage award
    Introduction, p-4
Wall coverings
    cotton backed vinyl, p-120
    decorative paper, p-119
    glass fibre, p-120
    suede effect, p-120
    textile wall coverings, p-119
    vinyl surfaced, p-120

Wallpaper
    decorative wall coverings, p-118
    heavy embossed, p-118
    lining paper, p-118
    sizing surfaces, p-118
    stripping off, p-137
Walls and ceilings
    burning off paintwork, p-130
    paint internally, p-30
    paint stripping, p-131
    acrylated rubber paint, p-41
Walls and ceilings, internally
    sprayed paint finishes, p-48
Walls externally
    masonry sealer, p-88
    paint externally, p-88
    Snowcem paint, p-89
    sprayed paint finishes, p-91
Water painted, lime washed surfaces
    redecoration, p-125, 126
Water repellent clear wood preserver on woodwork
    preservatives, p-109
Wood chip paper
    decorative wall coverings, p-118
Wood stain
    satin sheen - internally, p-81
Wood stain externally
    composite rates for painting, p-165
Woodwork
    burning off paintwork, p-135
    paint stripping, p-136
    redecoration, p-134
Woodwork - externally
    Cuprinol treatments, p-109
    Sadolin treatments, p-111
Woodwork externally
    composite rates for painting, p-163
    painting, p-104
Woodwork internally
    clear finish paint, p-81
    composite rates for painting, p-158

QSToolbox™
Weapons grade QS software

In these competitive times the systems you employ must keep you on top. There are no prizes for second place in business - coming first is the only option. Lose or win, sink or swim, it's your choice so make it the right one!

QSToolbox™ offers praise winning, profit making solutions for everyone. Drawing measurement - Bill Production - Estimating - Cost Planning, even dimsheet enhancement systems for your Excel™ spreadsheets.

If you want to move up a gear, call us now for your free DVD and brochure pack - it's free. Or go on-line and request your free DVD & brochure pack.

Less effort - less stress - more profits.

Come and see what the future holds.

Go on-line or call now to receive your free demonstration DVD and brochure pack.

Visual Precision

2 Sycamore Tree · Elmhurst Business Park
Park Lane · Elmhurst · Staffs · WS13 8EX
Tel: 01543 262 222 Fax: 01543 262 777
Email: sales@visualprecision.co.uk
Web: http://www.visualprecision.co.uk

CALL NOW | 01543 262 222   **WEB NOW | WWW.VISUALPRECISION.CO.UK**

INTRODUCTION

BCIS — 50 years celebrating excellence

CONSTRUCTION
BCIS PRICE DATA 2012

Comprehensive Building Price Book 2012
Major and Minor Works dataset

The Major Works dataset focuses predominantly on large 'new build' projects reflecting the economies of scale found in these forms of construction. The Minor Works Estimating Dataset focuses on small to medium sized 'new build' projects reflecting factors such as increase in costs brought about the reduced output, less discounts, increased carriage etc.

Item code: 18770
Price: £165.99

SMM7 Estimating Price Book 2012

This dataset concentrates predominantly on large 'new build' projects reflecting the economies of scale found in these forms of construction. The dataset is presented in SMM7 grouping and order in accordance with the Common Arrangement of Work Sections. New glazing section and manhole build-ups.

Item code: 18771
Price: £139.99

Alterations and Refurbishment Price Book 2012

This dataset focuses on small to medium sized projects, generally working within an existing building and reflecting the increase in costs brought about by a variety of factors including reduction in output, smaller discounts, increased carriage, increased supervision etc.

Item code: 18772
Price £109.99

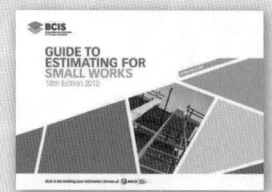

Guide to Estimating for Small Works 2012

This is a unique dataset which shows the true power of resource based estimating. A set of composite built-up measured items are used to build up priced estimates for a large number of common specification extensions.

Item code: 19064
Price: £59.99

For more information call **+44 (0)870 333 1600** email **contact@bcis.co.uk** or visit
www.bcis.co.uk/bcispricebooks

BCIS is the Building Cost Information Service of  RICS the mark of property professionalism worldwide

INTRODUCTION

INTRODUCTION

The aim of this book is to provide painting and decorating contractors and all those involved in the provision of estimates for painting and decorating work with a guide to current net cost unit prices at "national average" competitive rates. Pricing information is current as at December 2011 with labour rates based upon the national wage agreement effective from September 2011 (see under "Labour") and has been derived from the BCIS Database.

A further aim is to provide reference information of particular interest to the trade including the names and addresses of those associations, institutions and other bodies connected with the industry and useful technical information and conversion factors.

In calculating the labour hours required for the various items of work and in seeking prices for materials, it has been assumed that rates are for painting and decorating work up to a contract value of about £30,000. For very small works, allowance must be made to the rates for material purchases in small (and more expensive) quantities, possible increased wastage in part-used cans etc. and lack of continuity in labour utilisation, increase in transport and other overheads etc. Similarly, for very large painting and decorating works, the user should in particular anticipate more favourable trade discounts on bulk orders for materials, deliveries etc.

N.B. Prices <u>exclude</u> Value Added Tax

Prices are based upon site working conditions being reasonable - other trades being completed, or largely so, providing the painting and decorating contractor with proper access to the works, continuity of operation and all normal site facilities.

Descriptions of work and units are based upon the 6th edition of the Standard Method of Measurement of Building Works authorised by agreement between the Royal Institution of Chartered Surveyors and the Building Employers' Confederation.

Prices in the New Work Sections are given separately for first or priming coats (which include for preparation of the surfaces to be painted), undercoats and finishing coats. The total cost of decoration will be combinations of these individual coat prices in accordance with the required specification. A large number of permutations for various specifications is therefore possible. Refer to Composite Prices Section for examples of prices for composite painting and decorating items.

The undercoat and finishing coat prices should be similarly applied to the Redecoration Section which gives preparation prices and primers for the treatment of previously painted surfaces, again providing a large number of permutations for various redecorations which may be specified.

It must be stressed that the prices in this book are **guide prices** and cannot be guaranteed. Quotations for materials should be obtained for particular projects and detailed specifications, site working arrangements, scale and locations should be carefully taken into account by the user. In particular, regional variations in prices must be considered and, for an indication of the price differences, please refer to the section 'Regional Variations' below.

OVERHEADS AND PROFIT

The prices included in this book are net unit prices and **exclude** all overheads and profit which should be added by the user in accordance with the circumstances and management policy at the time of tender.

Allowance must be made by the user for all site overheads (e.g. plant, equipment, access scaffold, ladders, steps, dust sheets etc. and supervision, storage, messing facilities and services) as may be required in any particular project - as well as for off-site overheads (e.g. office, administration, insurances, transport, stores etc.). Brush and small tools allowances have been included in the prices (see section 'Brush Allowances').

INTRODUCTION

REGIONAL VARIATIONS

The prices generally are based upon nationally averaged 'best prices' with a factor of 1.00 .

Individual prices may be calculated by applying the relevant local rates obtained for labour and plant, to the hourly constants indicated against each item, together with the local cost of materials, using the methods described in this book.

An indication of the possible level of overall pricing in areas of the United Kingdom compared to the prices herein may be obtained by reference to the Regional Factors Map on the following page. 'Regional Factors' are taken from the BCIS Quarterly Review of Building Prices, reproduced by permission of its publishers, the Building Cost Information Service, Parliament Square, London, SW1P 3AD T: +44 (0)870 333 1600; F: +44 (0)20 7334 3851; E-Mail: contact@bcis.co.uk; Internet: www.bcis.co.uk

However, as the BCIS Quarterly Review is based upon total tender prices, it is stressed that this can only provide an approximate overall guide to the level of pricing and the figures in the guide should **never** be applied to individual prices under any circumstances.

LABOUR

Labour costs have been calculated in accordance with the recommendations of the Code of Estimating Practice published by the Chartered Institute of Building.

The calculations of hourly labour costs has been based on the Building and Allied Trades Joint Industry Council wage award and allowances payable to operatives from 12th September 2011.

The hourly rate calculations have been made to the nearest whole penny.

An enhanced 'plus rate' has been inserted in the calculation to suit local labour market conditions.

The hourly rates used for the calculations are therefore as follows:

Craftsman/Painter BATJIC	15.23	per hour
Labourer	11.25	per hour

NOTE: National Insurance rates have been based on those applicable from 6th April 2011

As travelling allowances vary with the distance of working sites from the contractors office, they have been excluded from the calculation of the hourly labour costs. Allowances for travelling time and/or expenses should be made in "overheads".

Trade supervision has been excluded from the calculation of the hourly labour costs. Where specific trade supervision is required, particularly on larger projects, separate allowances should be made in "overheads".

Allowance for overtime in the calculation of the hourly labour costs has been based on an average of 5 hours overtime per operative, per week, during British Summer Time. The non-productive element in the overtime amounts to an average of 65.5 hours per operative, per annum.

The labour constants, in the form of hours, against each item in the price book have been rounded off to two decimal places after calculation.

The labour constants are based on the assumption that operatives are working under average "motivated" conditions (implying incentive payments in some cases), the cost of which must be allowed for as necessary in the contractor's rates or in his "overheads", and that they are provided with reasonable working conditions, continuity of work and the tools and equipment suitable for the job.

INTRODUCTION

INTRODUCTION

Calculation of hours worked per annum:
Craftsmen and Labourers

	hours	hours	hours
(a) Summertime working: 30 weeks of British Summertime at 44 hours per week Monday to Friday			
30 Weeks at 44 hours		1320.0	
Less annual holidays (14 days)	123.2		
public holidays (5 days)	44.0	167.2	1152.8
	-----	------	
(b) Winter working: 22 weeks at 39 hours per week			
22 Weeks at 39 hours		858.0	
Less annual holidays (7 days)	54.6		
public holidays (3 days)	23.4		
Sick leave (8 days NB 3 days unpaid)	39.0	117.0	741.0
	-----	-----	------
Total Number of paid working hours during year			1893.8
Less Allowance for inclement weather (2%)			37.8

TOTAL NUMBER OF PRODUCTIVE HOURS WORKED PER ANNUM			1856.0

Calculation of Labour Costs - BATJIC award 2011

The hourly cost of wages based upon the rates of wages and allowances agreed by the Building and Allied Trades Joint Industry Council is calculated as follows:

Notes:
1. For trade supervision make allowance in "overheads".

2. For travelling expenses and allowances include in "overheads".

3. A 10p plus rate has been allowed and the effect of a further 10p variation is shown below the cost calculation.

4. The following rate has been applied in respect of craftsmen only. In practice on larger works labourers will be employed to assist craftsmen in preparation, cleaning etc. and allowance for the use of this unskilled labour must be made in appropriate cases.

5. Paperhangers have been included in the prices at craftsmen painter rates.

6. An all-in rate for spray painting has been calculated in the Basic Rates at the beginning of the Pricing Section and has been applied to the items specifically described as "sprayed". All other rates assume brush application, for which an allowance has been made as shown at the beginning of the Pricing Section. For adjustment of prices to allow for roller applied paint see the beginning of the Basic Rate Section.

7. The CITB advise a levy of 0.50% of PAYE payroll and 1.50% of labour-only sub-contract costs. The training allowance above was agreed in November 2000. Each variation of £20 per annum would equate to approximately 1p variation of the total hourly cost.

8. USERS OF THIS GUIDE SHOULD SATISFY THEMSELVES AS TO THE BASIS OF CALCULATION OF PAYMENTS TO OPERATIVES, RELATED COSTS, WORKING AND PRODUCTIVE TIME ETC., WITHIN THEIR OWN ORGANISATION AND REGION, BEFORE APPLYING ANY LABOUR RATES WITHIN THIS GUIDE. PARTICULAR ATTENTION IS DRAWN TO THE CHANGE IN RULES CONCERNING RECOVERY OF STATUTORY SICK PAY. ALTHOUGH NO ALLOWANCE HAS BEEN MADE IN THESE CALCULATIONS FOR RECOVERY OF SSP, IT SHOULD BE NOTED THAT THE RATES OF NATIONAL INSURANCE CONTRIBUTION WERE REDUCED FOR 1994/5 IN ORDER TO ALLOW FOR POSSIBLE INCREASED OVERALL SICKNESS COSTS, AND THAT THESE REDUCTIONS **ARE** INCORPORATED IN THE CALCULATION.

BCIS INTRODUCTION

9. With effect from 11th June 2001 the Holiday Credit and Retirement Benefit System ceased and was replaced with the Building and Civil Engineering Benefits Scheme. Holiday Pay Scheme - This sets out the percentage of PAYE to allow for holiday pay. The retirement benefit is accrued according to the stake-holders' pension scheme using the recommended allowance of £3.00 per week. However, if the employee chooses to pay a higher amount, up to £13.50, then this must be matched by the employer.

Annual cost of wages		Craftsman £		Labourer £
Flat time	1893.8 hours at 10.73	20320.47	at 7.96	15074.65
Non-productive overtime	65.5 hours at 10.73	702.82	at 7.96	521.38
Public holidays	63.0 hours at 10.73	675.99	at 7.96	501.48
Sick Pay	5.0 days at 23.61	118.05	at 23.61	118.05
Plus rate (See notes below)	2022.3 hours at 0.10	202.23	at 0.10	202.23
		22019.56		16417.79
Employer's National Insurance Contribution – Above ST	13.80%	2062.76	13.80%	1289.72
Training Allowance	(0.50% of PAYE)	110.10		82.09
Holiday credits	(12.6% of PAYE)	2774.46		2068.64
Retirement benefit	52.0 weeks at 3.00	156.00	at 3.00	156.00
Death benefit	12.0 months at 4.33	52.00	at 4.33	52.00
		27174.88		20066.24
Severance pay and other statutory costs	2.00%	543.50	2.00%	401.32
		27718.38		20467.56
Employer's liability insurance	2.00%	554.37	2.00%	409.35
TOTAL COST OF 1856 PRODUCTIVE HOURS		28272.75		20876.91
Total Labour Cost per hour		**15.23**		**11.25**
Effect of 10p/hr plus rate on total cost per hour (see note 2 below)		0.1439		0.1439

MATERIALS

Prices of materials are current as at the fourth quarter of 2011 for quantities as previously described in page 1 and in packs stated in the Unit column of the Basic Prices Section.

An allowance for wastage has been made and shown as a percentage against each material listed to cover normal waste-in-use and spillage.

The supply prices of materials are Trade prices for goods normally supplied and collected through Builders' Merchants. No allowance has been made for special, quantity, cash or other discounts which may be obtained by individual arrangement.

In the description of materials, particular brand names have been avoided where possible but prices are based on the assumption that good quality products from reputable manufacturers will be used throughout. Where a given material has a wide variation in price (e.g. wallpapers) P.C. prices are given and these should be substituted by the cost of the actual material; specified for the work.

The traditional "oil paints" have been replaced by manufacturers with alkyd-based paints and the material has therefore been described as such in this book. However, the term "oil paint" is still in common use in the trade and indeed in some specifications and, in such cases, readers should refer to the alkyd-based prices given.

No thinning of paint has been allowed for in the prices unless specifically described herein or called for by the manufacturers own recommendations (e.g. for spray application) and the Net Materials price is for neat paints from the can.

RIPAC

The total cost control and contract administration system that maximises performance and flexibility from minimised input

QS's, Consulting Engineers, Contractors, Project Managers, Developers and Client Bodies

- Budget estimates
- Cost planning
- Bills of quantities
- Measurement from CAD / BIM
- Tender pricing and appraisal
- E-tendering
- Resource analysis
- Programme planning links
- Post contract administration
- Payments
- Cash flow
- Whole life costing

Integrated cost control through all project stages.
Easy manipulation of data.
Outputs in user defined formats.
Speedy revisions and updates.
Previous projects available for re-use, analysis and benchmarking.

Various standard libraries of descriptions and price data bases including  **BCIS** Independent cost information for the built environment

www.cssp.co.uk

29 London Road
Bromley Kent BR1 1DG
Tel 020 8460 0022
Fax 020 8460 1196
Email enq@cssp.co.uk

GENERAL INFORMATION

BCIS 50 years celebrating excellence

CONSTRUCTION • MAINTENANCE COSTS & REPAIRS • REBUILDING COSTS • INTELLIGENCE

2012

HOW LONG? HOW MUCH?
THE FASTEST, MOST UP-TO-DATE ANSWERS ARE AVAILABLE NOW

Cost information underpins every aspect of the built environment, from construction and rebuilding to maintenance and operation publications.

BCIS, the RICS' Building Cost Information Service, is the leading provider of cost information to the construction industry and anyone else who needs comprehensive, accurate and independent data.

For the past 50 years, BCIS has been collecting, collating, analysing, modelling and interpreting cost information. Today, BCIS make that information easily accessible through online applications, data licensing and publications.

For more information call **+44 (0)870 333 1600** email **contact@bcis.co.uk** or visit **www.bcis.co.uk**

BCIS is the Building Cost Information Service of  RICS the mark of property professionalism worldwide

GENERAL INFORMATION

General Information — Average coverage of paint per coat

The following table is reproduced by permission of: The Paint and Painting Industries' Liaison Committee - Constituent bodies: - British Decorators Association, National Federation of Painting and Decorating Contractors, Paintmakers Association of Great Britain and Scottish Decorators Federation.

The schedule of average coverage figures in respect of painting work is the 1974 revision (with amendments as at April 2000) of the schedule compiled and approved for the guidance of commercial organisations and professional bodies when assessing the values of materials in painting work

In this revision a range of spreading capacities is given. Figures are in square metres per litre, except for oil-bound water paint and cement-based paint which are given in square metres per kilogram.

For comparative purposes figures are given for a single coat, but users are advised to follow manufacturers' recommendations as to when to use single or multicoat systems.

It is emphasised that the figures quoted in the schedule are practical figures for brush application, achieved in scale painting work and take into account losses and wastage. They are not optimum figures based upon ideal conditions of surface, nor minimum figures reflecting the reverse of these conditions.

There will be instances when the figures indicated by paint manufacturers in their literature will be higher than those shown in the schedule. The committee realise that under ideal conditions of application, and depending on such factors as the skill of the applicator and the type and quality of the product, better covering figures can be achieved.

The figures given below are for application by brush and to appropriate systems on each surface. They are given for guidance and are qualified to allow for variation depending on certain factors.

Type of surface

Coating (m2 per litre)	Finishing plaster	Wood floated rendering	Smooth concrete /cement	Fair faced brickwork	Fair faced blockwork	Roughcast/ Pebbledash	Hard board	Soft fibre insulating board
Water thinned primer/undercoat								
as primer	13-15	-	-	-	-	-	10-12	7-10
as undercoat	-	-	-	-	-	-	-	7-10
Plaster primer (including building board)	9-11	8-12	9-11	7-9	5-7	2-4	8-10	7-9
Alkali resistant primer	7-11	6-8	7-11	6-8	4-6	2-4		
External wall primer sealer	6-8	6-7	6-8	5-7	4-6	2-4		
Undercoat	11-14	7-9	7-9	6-8	6-8	3-4	11-14	7-10
Gloss finish	11-14	8-10	8-10	7-9	6-8	-	11-14	7-10
Eggshell /semi-gloss finish (oil based)	11-14	9-11	11-14	8-10	7-9	-	10-13	7-10
Emulsion paint:								
Standard	12-15	8-12	11-14	8-12	6-10	2-4	12-15	8-10
Contract	10-12	7-11	10-12	7-10	5-9	2-4	10-12	7-9
Heavy textured coating	2-4	2-4	2-4	2-4	2-4	-	2-4	2-4
Masonry paint	5-7	4-6	5-7	4-6	3-5	2-4	-	-
Cement based paint	-	4-6	6-7	3-6	3-6	2-3	-	-

GENERAL INFORMATION

Type of surface Coating (m2 per litre)	Fire retardent fibre insulating board	Smooth paper faced board	Hard asbestos sheet	Structural steelwork	Metal sheeting	Joinery	Smooth primed surfaces	Smooth Under-coated surfaces
Wood primer (oil based)	-	-	-	-	-	8-11	-	-
Water thinned primer undercoat								
as primer	-	8-11	7-10	-	-	10-14	-	-
as undercoat	-	10-12	-	-	-	12-15	12-15	-
Aluminium sealer:*								
spirit based	-	-	-	-	-	7-9	-	-
oil based	-	-	-	-	9-13	9-13	-	-
Metal primer:								
Conventional	-	-	-	7-10	10-13	-	-	-
Plaster primer (including building board)	8-10	10-12	10-12	-	-	-	-	-
Alkali resistant primer	-	-	8-10	-	-	-	-	-
External wall primer sealer	-	-	6-8	-	-	-	-	-
Undercoat	10-12	11-14	10-12	10-12	10-12	10-12	11-14	-
Gloss finish	10-12	11-14	10-12	10-12	10-12	10-12	11-14	11-14
Eggshell/ semi-gloss finish (oil based)	10-12	11-14	10-12	10-12	10-12	10-12	11-14	11-14
Emulsion paint:								
Standard	8-10	12-15	10-12	-	-	10-12	12-15	12-15
Contract	-	10-12	8-10	-	-	10-12	10-12	10-12
Heavy textured coating	2-4	2-4	2-4	2-4	2-4	2-4	2-4	2-4
Masonry paint	-	-	5-7	-	-	-	8-10	6-8
Cement based paint	-	-	4-6	-	-	-	-	-

NOTES to preceding tables

*Aluminium primer/sealer is normally used over 'bitumen' painted surfaces.

Specialised metal primer: Figures should be obtained from individual manufacturers

Oil based thixotropic finish: Figures should be obtained from individual manufacturers

Glossy emulsion: Figures should be obtained from individual manufacturers

The texture of roughcast, Tyrolean and pebbledash can vary markedly and thus there can be significant variations in the coverage of paints applied to such surfaces. The figures given are thought to be typical but under some circumstances much lower coverage will be obtained.

In many instances the coverages achieved will be affected by the suction and texture of the backing; for example, the suction and texture of brickwork can vary to such an extent that coverages outside those quoted may be on occasions obtained.

GENERAL INFORMATION

It is necessary to take these factors into account when using this table.

Brush Allowance

Allowance has been made in the 'Net Materials' prices for brushes, rollers, pads, sandpaper, pumice stones and the like accordance with the following table:

SURFACE	Allowance made per coat per m2 £
V0005 Plaster	0.02
V0006 Smooth concrete	0.03
V0007 Asbestos cement	0.02
V0008 Textured paper	0.02
V0009 Cement render	0.02
V0010 Fair face brick	0.03
V0011 Fair face blockwork	0.04
V0012 Smooth timber	0.02
V0013 Sawn timber	0.03
V0014 Metalwork	0.02
V0015 Pebbledash	0.04
V0016 Rough cast	0.03
V0017 Tyrolean render	0.06
V0018 Plasterboard	0.02
V0019 Paperhanging (generally)	0.02

PREPARATION FOR PAINTING

The preparation of the surfaces to be painted is often crucial to the success or otherwise of the paint film and specifiers attach the utmost importance to this item. It is also extremely labour intensive and therefore has a major impact on the labour costs and ultimately the Net Unit Prices. In most cases throughout this book the task of preparation is described as "Prepare" and this should be interpreted for different surfaces as follows:-

Woodwork

Rub down with glass paper to a smooth surface finishing with fine grade paper. Lightly round off sharp edges of timber with the glass paper. Fill surface blemishes with patent stopping or filler finished smooth. Finally, dust off surfaces to remove all loose material before painting. The wood should be dry before paint is applied.

Metalwork

Ferrous metals (iron and steel) should be wire brushed, scraped and smoothed with fine emery paper to remove all rust and scale and other corrosion and thoroughly washed down with white spirit to remove grease, oil and dust. Ideally, surfaces should be shot blasted or grit blasted in preference to wire brushing and scraping but such treatment is normally separately called for in the specification. (Painting treatment should follow preparation as quickly as possible to prevent further surface corrosion).

Non-ferrous metals (aluminium, copper, zinc etc.) generally require only light preparation with fine emery paper but it is essential to thoroughly clean down surfaces with white spirit to remove all oil and grease before painting with special primers.

Galvanised surfaces will require cleaning to remove grease etc. and coating with mordant solution before painting.

Plasterwork

New plasterwork must be allowed to dry out thoroughly before any paint finish is applied. Brush down surfaces to remove any dusting or wipe over surface with a damp cloth, stop-in any surface blemishes and lightly rub down any nibs.

GENERAL INFORMATION

Cement Rendering

Brush down surfaces to remove dusting and stop-in surface blemishes. The rendering must be thoroughly dried-out before paint finishes are applied.

Asbestos Cement

Caution must be exercised in the preparation of these surfaces because of the health risks associated with asbestos fibres. Care must be taken to ensure that no fibres or dust are released from the surface and hence any preparation must be a "wet" treatment. Material released must be collected into suitable plastic bags or containers, sealed, labelled and safely disposed of after consultation with the safety officer.

Generally, preparation should consist of lightly rubbing down with wet emery paper and cleaning down with damp cloths. The surfaces should be allowed to dry before painting.

Brickwork, Blockwork etc. and Concrete

Brush down surfaces to remove loose material, dusting, cement splashes etc. Surfaces should have dried out thoroughly before painting.

Previously painted surfaces

Scrape off all loose, blistered or flaky paint, rub down with glass paper feathering off edges of paint film, clean down with white spirit to remove all oil and grease. Stop-in surface blemishes. Bare patches should be brought forward in the specified paints to a smooth and even surface for new paint work. Paintwork in a bad condition may require complete removal by burning-off or by chemical stripping. Both processes are highly labour intensive and both require considerable preparation (rubbing down, smoothing, filling, washing-off etc.) before applying new paint.

Preparation between coats

NB: Labour prices include an allowance for light rubbing down, dusting/wiping down between coats of paint in order to achieve a good standard of finish to the final coat.

Further technical information

Where there is no specification for the painting and decorating work or when particular problems associated with the trade are encountered, advice can usually be obtained from the technical department of the major paint manufacturers or from one of the bodies included in the list given elsewhere in this book.

GENERAL INFORMATION

HEALTH AND SAFETY

There are several hazards involved in the carrying out of painting and decorating work which are directly connected with the materials and methodology employed in addition to those encountered in the building industry in general: 1. inhaling fumes or vapours, 2. contact of materials with skin or eyes and 3. fire risk are the main special hazards associated with painting and decorating. The Health and Safety at Work Act 1974 requires manufacturers and suppliers to draw users attention to potential hazards of their products and to recommend precautions or methods for their use to minimise risk. Users should therefore take careful note of manufacturers' instructions for the use of their particular products.

General points to note:

1. Fumes/vapours

Work areas should be well ventilated. Operatives should be provided with masks if necessary. smoking should not be permitted. Cans, bottles etc. should be resealed when not in use. Food and drink should not be stored or consumed in the work area. Application by spray may require forced ventilation and respirators to be issued to the operatives.

2. Contact with the skin

Some people are very susceptible to skin irritation by the chemicals in paints and thinners etc. and protective clothing, including suitable gloves and eye protectors, should be provided. Barrier creams should be available to operatives. Easy access to a supply of clean water is desirable for douching/rinsing splashes from the skin and eyes.

3. Fire risk

Nearly all the materials used in painting and decorating are flammable. Flame spread can be rapid and dense, noxious smoke/fumes are given off. Materials should therefore be stored in a safe place away from possible sources of ignition. Minimum quantities should be held on site. Lids and caps should be firmly in place. Foam or dry powder fire extinguishers should be accessible to the storage area.

Safety precautions regarding safe access in the construction industry apply equally to painting and decorating. Ladders, steps, trestles, planks etc. should be kept in good condition and inspected regularly for defects. Any defective access equipment should be immediately removed from site and if beyond repair should be destroyed or disposed of.

Access equipment should be used from a firm, level base and should be suitably secured top and bottom.

An information booklet covering most Health and Safety aspects of painting and decorating is produced by the National Federation of Painting and Decorating Contractors.

BCIS 50 years celebrating excellence

CONSTRUCTION

LESS DESK TIME MORE FREE TIME

THE REVOLUTION OF THE PRICE BOOK IS HERE
BCIS ONLINE RATES DATABASE

As a purchaser of the 2012 price book, we would like to offer you a free two week trial of our Online Rates Database. It's the price book but online, with lots of additional features to help you to:

- locate prices quickly using faster navigation
- adjust your data to suit a time frame of your choice and choose location factors to make your costs more accurate.

Plus:

- everything you need is in one place – you have a full library at your disposal
- with the full service you can download data into an Excel spreadsheet and manipulate and store your data electronically*.

*During the two week trial data downloads are not available

Offering immediate online access to independent BCIS resource rates data, quantity surveyors and others in the construction industry can have all the information needed to compile and check estimates on their desktops. You won't need to worry about being able to lay your hands on the office copy of the latest price books, all of the information is now easily accessible online.

For a **FREE TRIAL** of BCIS online rates database, register at **www.bcis.co.uk/ordbdemo**

- Accuracy
- Futureproof
- Value for money
- Saves time

- Flexible
- Customise
- Portable
- Comprehensive

BCIS is the Building Cost Information Service of  RICS the mark of property professionalism worldwide

BASIC MATERIAL PRICES

RICSBOOKS.COM

Leading supplier of books and contracts for the surveying, construction and property profession.

Shop online at www.ricsbooks.com

- Free delivery on orders over £75
- Exclusive web offers & discounts
- Standard & next day delivery
- Open when you want 24/7

BASIC RATES

This Section shows the hourly rates of pay for operatives and the supply prices of materials (with an allowance for "waste") which have been used for compilation of the basic prices included in the subsequent Unit Prices and Composite Prices sections.

The detailed calculations used in arriving at the hourly rates of pay for operatives are given in the Introduction Section.

isurv

Building value from knowledge

RICS — the mark of property professionalism worldwide

200 forms, 4000 case studies, unlimited guidance

Yours in a couple of clicks

From Building Surveying to Planning and Valuation. From Commercial Property to Construction and Environment. From Estate Agency to Legal to Sustainability.

isurv gives you the most complete online RICS guidance, ever.

With detailed information on best practice, legal guidance, compliance, case summaries, legislative updates, expert commentary and over 200 downloadable forms; isurv is the definitive source for all property professionals.

And, with more content and more channels, isurv now costs less-per-channel than ever before. Can you afford not to take our free trial?

FREE 7-day trial
Go to **www.isurv.com** or call **024 7686 8433** now

BASIC RATES

Labour Rates

Artexer
   Artexer - £15.23 per hour

Paint sprayer
   Paint Sprayer - £15.23 per hour

Painter and decorator
   Painter and Decorator - £15.23 per hour

PAPERHANGING, sheet plastic and fabric linings
   Craftsman (BATJIC Craft Rate) - £15.23 per hour

SPRAY PAINTING, based upon the cost of labour plus spraying equipment calculated as follows:

COST PER WEEK	£
Hire of (2) gun electric spray equipment	180.00
Use of face masks etc.	4.30
Masking materials, tape etc.	6.50
Cleaning fluids	4.50
Painters (2) x £15.23 per hour x 39 hrs	1187.94
Total weekly cost	**£1383.24**

ACTUAL HOURS WORKED PER WEEK		Hours
(2) Painters at 39 hours		78 hrs
Less cleaning up time (half hr per day)	5 hrs	
masking time (1.5 hours per day)	15 hrs	20 hrs
NET hours worked per week		58 hrs

COST PER HOUR

Total weekly cost divided by Net Hours worked (£1383.24 divided by 58)

Cost per hour   £ 23.85

Adjustment for roller applied paints
The labour and material content of unit rates within this section are based upon paints applied by brush. For guidance to the cost of painting using a roller, the following percentage adjustments, obtained from industry specialists, should be applied:
a)   Net Labour Price   -   Reduce by 15%
b)   Net Material Price   -   Increase by 7.5%

Brush Allowance
Allowance has been made in the 'Net Materials' prices for brushes, rollers, pads, sandpaper, pumice stones and the like in accordance with the following table:

Surface	Allowance made per coat per m2 £
V0005 Plaster	0.02
V0006 Smooth concrete	0.03
V0007 Fibre cement	0.02
V0008 Textured paper	0.02
V0009 Cement render	0.02
V0010 Fair face brick	0.03
V0011 Fair face blockwork	0.04
V0012 Smooth timber	0.02
V0013 Sawn timber	0.03
V0014 Metalwork	0.02
V0015 Pebbledash	0.04
V0016 Rough cast	0.03
V0017 Tyrolean render	0.06
V0018 Plasterboard	0.02
V0019 Paperhanging (generally)	0.02

Thinners
Unless specifically stated in the descriptions, no thinning of paint has been allowed and the 'Net Materials' price is for neat paints.

Basic Prices of Materials

BASIC RATES

		Supply Price £	Waste Factor %	Unload. Labour £	Unload. Plant £	Total Unit Cost £	Unit
	BASIC PRICES OF MATERIALS						
	PAINTS						
	Primers:						
V0101	all purpose primer	14.70	2.5%	-	-	15.07	ltr
V0102	wood primer, white and pink	6.96	2.5%	-	-	7.13	ltr
V0103	metal primer, zinc phosphate	7.67	2.5%	-	-	7.86	ltr
V0104	metal primer, zinc chromate	7.44	2.5%	-	-	7.62	ltr
V0105	metal primer, zinc rich	37.50	2.5%	-	-	38.43	ltr
V0106	metal primer, red oxide	7.63	2.5%	-	-	7.83	ltr
V0107	alkali resisting primer	12.19	2.5%	-	-	12.50	ltr
V0108	aluminium wood primer	13.23	2.5%	-	-	13.56	ltr
V0109	acrylic primer undercoat (white)	6.96	2.5%	-	-	7.13	ltr
V0110	ICI quick drying wood primer	11.88	2.5%	-	-	12.18	ltr
V0111	etching primer (2 pack)	20.00	2.5%	-	-	20.50	ltr
V0112	thinners for etching primer	11.18	2.5%	-	-	11.45	ltr
V0113	preservative primer	11.70	2.5%	-	-	11.99	ltr
V0114	WSE primer / adhesion promoter high performance water-based anti bacterial, two component epoxy coating	14.38	7.5%	-	-	15.46	ltr
	Undercoats:						
V0120	white	5.82	2.5%	-	-	5.97	ltr
V0121	brilliant white	5.82	2.5%	-	-	5.97	ltr
V0122	standard colours	6.41	2.5%	-	-	6.57	ltr
	Eggshell paints:						
V0130	white	6.36	2.5%	-	-	6.51	ltr
V0131	brilliant white	6.36	2.5%	-	-	6.51	ltr
V0132	standard colours	7.00	2.5%	-	-	7.17	ltr
	Gloss paints:						
V0140	white	5.82	2.5%	-	-	5.97	ltr
V0141	brilliant white	6.25	2.5%	-	-	6.41	ltr
V0142	standard colours	6.41	2.5%	-	-	6.57	ltr
	Emulsion paints:						
	silk:						
V0150	white	5.30	2.5%	-	-	5.44	ltr
V0151	brilliant white	5.53	2.5%	-	-	5.67	ltr
V0152	standard colours	5.82	2.5%	-	-	5.97	ltr
	matt:						
V0153	white	4.58	2.5%	-	-	4.69	ltr
V0154	brilliant white	4.84	2.5%	-	-	4.96	ltr
V0155	standard colours	8.62	2.5%	-	-	8.83	ltr
	Metal finish:						
V0160	Hammerite 'hammered' metal finish	14.56	2.5%	-	-	14.93	ltr
V0161	Hammerite 'smooth' metal finish	14.56	2.5%	-	-	14.93	ltr
V0162	Hammerite No. 1 metal primer	12.76	2.5%	-	-	13.08	ltr
V0163	Hammerite thinners	6.70	2.5%	-	-	6.87	ltr

BASIC RATES

Basic Prices of Materials

Code	Description	Supply Price £	Waste Factor %	Unload. Labour £	Unload. Plant £	Total Unit Cost £	Unit
	Acrylated rubber paints:						
V0170	finish	18.32	2.5%	-	-	18.78	ltr
V0171	finish, rich colours	23.21	2.5%	-	-	23.79	ltr
V0172	thick coating brushing	19.43	2.5%	-	-	19.92	ltr
V0173	thick coating airless spray	17.63	2.5%	-	-	18.08	ltr
V0174	primer undercoat, off white	18.76	2.5%	-	-	19.23	ltr
V0175	metal primer	12.96	2.5%	-	-	13.29	ltr
V0176	acrylated thinners	7.08	2.5%	-	-	7.26	ltr
	Fire resisting paints:						
V0401	Thermoguard Thermocoat intumescent paint - high build primer	14.26	7.5%	-	-	15.33	ltr
V0402	Thermoguard Thermocoat intumescent paint - base coat	14.26	7.5%	-	-	15.33	ltr
V0403	Thermoguard Thermocoat intumescent paint - top coat	14.26	7.5%	-	-	15.33	ltr
V0404	Thermoguard Wallcoat - intumescent paint - top coat	18.34	7.5%	-	-	19.71	ltr
V0405	Thermoguard Timbercoat - intumescent paint	18.34	7.5%	-	-	19.71	ltr
V0406	Thermoguard Fire Varnish - intumescent varnish - coverage 6m2/1	14.26	7.5%	-	-	15.33	ltr
V0407	Thermoguard sealer for fire varnish basecoat. Available in matt, stain and gloss	16.17	7.5%	-	-	17.38	ltr
V0408	Thermoguard flame retardant acrylic matt	14.26	7.5%	-	-	15.33	ltr
	Flooring Paints:						
V0166	non slip floor paint with bauxite aggregate	5.19	5%	-	-	5.45	ltr
V0167	Bartoline creosote d/brown 201	1.67	5%	-	-	1.75	ltr
V0168	quick drying Thermoplastic protective floor paint	4.19	5%	-	-	4.40	ltr
V0243	line marking paint - chlorinated rubber	7.61	7.5%	-	-	8.18	ltr
V0244	line paint thinners (th14) (1:10)	7.43	7.5%	-	-	7.99	ltr
	Sundries:						
V0178	fungicidal solution	3.31	5%	-	-	3.48	ltr
V0179	paint stripper	6.75	5%	-	-	7.09	ltr
V0180	masonry stabilising sealer	8.60	5%	-	-	9.03	ltr
V0181	aluminium paint	13.35	2.5%	-	-	13.68	ltr
V0182	floor paint, tile red	6.18	2.5%	-	-	6.34	ltr
V0183	'Snowcem' cement based paint (white)	1.28	5%	-	-	1.34	kg
V0191	patent knotting	28.91	10%	-	-	31.80	ltr
V0185	boiled linseed oil	4.51	5%	-	-	4.74	ltr
V0186	raw linseed oil	3.97	5%	-	-	4.17	ltr
V0187	Interior clear varnish (gloss, satin or matt)	11.39	2.5%	-	-	11.68	ltr
V0188	'Artex' AX compound	0.71	5%	-	-	0.75	kg
V0189	'Artex' joint tape 64 mm wide 180 m roll	9.70	10%	-	-	10.67	nr
V0190	'Artex' sealer	4.90	10%	-	-	5.39	ltr
V0240	Fleck Finish topcoat colours	9.58	2.5%	-	-	9.82	ltr
V0241	Fleck-Stone basecoat colours	8.13	2.5%	-	-	8.34	ltr
V0242	High Durability Clearcoat gloss	9.23	2.5%	-	-	9.46	ltr
V0192	masonry paint, brilliant white, smooth	8.36	10%	-	-	9.20	ltr
V0193	'Sandtex' Exterior Matt (white)	3.39	10%	-	-	3.73	ltr
V0194	masonry stabilising solution	6.04	10%	-	-	6.64	ltr
V0195	'Sandtex' Fine Build	2.66	10%	-	-	2.93	kg
V0196	'Sandtex' High Build	3.26	10%	-	-	3.58	kg
V0197	Fosroc 'Galvafroid' galvanising paint	35.73	5%	-	-	37.51	ltr
V0198	Exterior varnish	15.87	2.5%	-	-	16.27	ltr

BASIC RATES

Basic Prices of Materials

Code	Description	Supply Price £	Waste Factor %	Unload. Labour £	Unload. Plant £	Total Unit Cost £	Unit
V0199	White spirit	3.57	5%	-	-	3.75	ltr
V0381	Aquagene hygiene paint water-based, two pack polyurethane system (8 m2/litre per coat)	20.78	7.5%	-	-	22.34	ltr
V0391	white peak Buxton limewash - white	2.01	7.5%	-	-	2.16	ltr
	The following prices were supplied by ICI Paints;						
	Wood treatments:						
	Wood preservers:						
V0200	clear wood preserver	5.35	5%	-	-	5.61	ltr
V0201	light oak/dark oak preserver	5.35	5%	-	-	5.61	ltr
V0202	green wood preserver	5.35	5%	-	-	5.61	ltr
V0203	exterior wood preserver	3.27	5%	-	-	3.44	ltr
V0204	Ducksback	2.45	5%	-	-	2.57	ltr
	Fencing treatments:						
V0203	exterior wood preserver	3.27	5%	-	-	3.44	ltr
V0205	Garden Timbercare	1.73	5%	-	-	1.81	ltr
V0206	Garden Shades	5.64	5%	-	-	5.92	ltr
V0207	Landscape Shades	13.92	5%	-	-	14.61	ltr
V0208	Landscape Stain	12.24	5%	-	-	12.85	ltr
V0210	decorative preserver (i.e. red cedar)	6.23	5%	-	-	6.55	ltr
	Wood stains:						
V0211	Premier 5	14.18	5%	-	-	14.89	ltr
V0212	Select	13.59	5%	-	-	14.27	ltr
	Interior floor finishes:						
V0217	Original Bourneseal (gloss)	12.58	2.5%	-	-	12.90	ltr
V0218	Quick drying Bourneseal (gloss/satin)	12.34	2.5%	-	-	12.65	ltr
	Varnishes:						
V0213	Trade polyurethane clear (gloss/satin/matt)	8.78	2.5%	-	-	9.00	ltr
V0214	Trade Quick Dry clear (gloss/satin)	10.28	2.5%	-	-	10.54	ltr
V0216	Trade Quick Dry colours (satin)	12.35	2.5%	-	-	12.66	ltr
V0215	Yacht varnish (exterior - gloss)	9.28	2.5%	-	-	9.51	ltr
	The following prices are indicative for Sadolin products:						
	Wood preservers and stains:						
V0230	'Classic' clear base	10.16	5%	-	-	10.67	ltr
V0231	'Classic' decorative timber protection	10.16	5%	-	-	10.67	ltr
V0232	'Superdec' opaque wood protection	11.66	5%	-	-	12.24	ltr
V0233	'Extra' joinery protection	10.16	5%	-	-	10.67	ltr
V0237	Hardwood Finish	10.08	2.5%	-	-	10.33	ltr
V0238	Breather Paint	10.54	5%	-	-	11.06	ltr

BASIC RATES

Basic Prices of Materials

Code	Description	Supply Price £	Waste Factor %	Unload. Labour £	Unload. Plant £	Total Unit Cost £	Unit
	Decorative paper and other wall coverings						
	The prices for wallpaper are per standard roll of 5.3 m2, however due to the varying sizes of other wall coverings these are priced per square metre.						
	Waste has been allowed in the constants used for calculating the 'Net Materials' prices. This waste factor may increase for pattern match materials. Check with manufacturer's literature for pattern repeats.						
	Prices have been averaged for the various types and are offered as P.C. sums per roll, except for textile coverings which are given as P.C. prices per square metre.						
V0250	lining paper	1.98	5%	-	-	2.07	nr
V0251	woodchip paper	1.98	-	-	-	1.98	nr
V0252	vinyl surfaced paper	11.44	10%	-	-	12.58	nr
V0253	ready pasted vinyl surfaced paper	14.56	10%	-	-	16.02	nr
V0254	mid-range quality paper	12.48	10%	-	-	13.73	nr
V0255	heavy embossed paper for painting	9.36	10%	-	-	10.30	nr
V0275	blown vinyl surface paper	7.79	10%	-	-	8.57	nr
V0350	lining paper P.C.	1.90	-	-	-	1.90	nr
V0351	woodchip paper P.C.	1.90	-	-	-	1.90	nr
V0352	vinyl surfaced paper P.C.	11.00	-	-	-	11.00	nr
V0353	ready pasted vinyl surfaced paper P.C.	14.00	-	-	-	14.00	nr
V0354	mid-range quality paper P.C.	12.00	-	-	-	12.00	nr
V0355	heavy embossed paper for painting P.C.	9.00	-	-	-	9.00	nr
V0375	blown vinyl surface paper P.C.	7.49	-	-	-	7.49	nr
	Wallpaper paste:						
V0256	ordinary	0.50	10%	-	-	0.55	ltr
V0257	heavy duty	0.67	10%	-	-	0.73	ltr
V0273	size	2.24	10%	-	-	2.47	ltr
	Other wall coverings: (Wallcoverings i.e. wide width contract vinyls are supplied in 30 - 50 m rolls, but other lengths are available. Due to varying sizes, prices are quoted per m2; compiled from information given by Muraspec Ltd)						
V0258	high quality textile	30.25	10%	-	-	33.28	m2
V0259	mid-range textile	22.66	10%	-	-	24.93	m2
V0260	glass fibre	5.90	10%	-	-	6.49	m2
V0261	suede effect	23.98	10%	-	-	26.38	m2
V0262	fabric backed vinyl	11.00	10%	-	-	12.10	m2
V0263	paper backed vinyl	2.40	10%	-	-	2.64	m2
V0276	hessian wall covering	6.24	10%	-	-	6.86	m2
V0358	high quality textile P.C.	27.50	-	-	-	27.50	m2
V0359	mid-range textile P.C.	20.60	-	-	-	20.60	m2
V0360	glass fibre P.C.	5.36	-	-	-	5.36	m2
V0361	suede effect P.C.	21.80	-	-	-	21.80	m2
V0362	fabric backed vinyl P.C.	10.00	-	-	-	10.00	m2
V0363	paper backed vinyl P.C.	2.18	-	-	-	2.18	m2

BASIC RATES

Basic Prices of Materials

		Supply Price £	Waste Factor %	Unload. Labour £	Unload. Plant £	Total Unit Cost £	Unit
V0269	Standard adhesive	3.02	10%	-	-	**3.33**	kg
V0270	Grade 1 adhesive (heavy)	3.39	10%	-	-	**3.73**	kg
V0271	Grade 2 adhesive (medium)	3.02	10%	-	-	**3.33**	kg
V0272	Grade 3 adhesive (light)	1.99	10%	-	-	**2.19**	kg

BCIS — 50 years celebrating excellence

BCIS PRICE DATA 2012 — CONSTRUCTION

Comprehensive Building Price Book 2012
Major and Minor Works dataset

The Major Works dataset focuses predominantly on large 'new build' projects reflecting the economies of scale found in these forms of construction. The Minor Works Estimating Dataset focuses on small to medium sized 'new build' projects reflecting factors such as increase in costs brought about the reduced output, less discounts, increased carriage etc.

Item code: 18770
Price: £165.99

SMM7 Estimating Price Book 2012

This dataset concentrates predominantly on large 'new build' projects reflecting the economies of scale found in these forms of construction. The dataset is presented in SMM7 grouping and order in accordance with the Common Arrangement of Work Sections. New glazing section and manhole build-ups.

Item code: 18771
Price: £139.99

Alterations and Refurbishment Price Book 2012

This dataset focuses on small to medium sized projects, generally working within an existing building and reflecting the increase in costs brought about by a variety of factors including reduction in output, smaller discounts, increased carriage, increased supervision etc.

Item code: 18772
Price £109.99

Guide to Estimating for Small Works 2012

This is a unique dataset which shows the true power of resource based estimating. A set of composite built-up measured items are used to build up priced estimates for a large number of common specification extensions.

Item code: 19064
Price: £59.99

For more information call **+44 (0)870 333 1600** email **contact@bcis.co.uk** or visit
www.bcis.co.uk/bcispricebooks

BCIS is the Building Cost Information Service of **RICS** — the mark of property professionalism worldwide

NEW WORK INTERNALLY

BCIS 50 years celebrating excellence

CONSTRUCTION · MAINTENANCE COSTS & REPAIRS · REBUILDING COSTS · INTELLIGENCE

2012

HOW LONG? HOW MUCH?
THE FASTEST, MOST UP-TO-DATE ANSWERS ARE AVAILABLE NOW

Cost information underpins every aspect of the built environment, from construction and rebuilding to maintenance and operation publications.

BCIS, the RICS' Building Cost Information Service, is the leading provider of cost information to the construction industry and anyone else who needs comprehensive, accurate and independent data.

For the past 50 years, BCIS has been collecting, collating, analysing, modelling and interpreting cost information. Today, BCIS make that information easily accessible through online applications, data licensing and publications.

For more information call **+44 (0)870 333 1600** email **contact@bcis.co.uk** or visit **www.bcis.co.uk**

BCIS is the Building Cost Information Service of RICS — the mark of property professionalism worldwide

INTERNAL WORK

NEW WORK INTERNALLY

Prices in this section are given separately for first or priming coats (which include for preparation of the surfaces to be painted), undercoats and finishing coats. The total cost of decoration will be combinations of these individual coat prices in accordance with the required specification.

A large number of permutations for various specifications is therefore possible.

Refer to the Composite Prices section for examples of prices for composite painting and decorating items.

Unit Rates

		Man-Hours	Net Labour Price £	Net Mats Price £	Net Unit Price £	Unit
VA	**UNIT RATES** NEW WORK INTERNALLY (SEE ALSO SECTION VZ FOR COMPOSITE EXAMPLES)					
	Prices are given separately for first or priming coats (which include for preparation of the surfaces to be painted), undercoats and finishing coats. The total cost of decoration will be combinations of these individual coat prices in accordance with the required specification. A large number of permutations for various specifications is therefore possible. Refer to Section VZ for examples of prices for composite painting and decorating items					
	WALLS AND CEILINGS - INTERNALLY					
	EMULSION PAINT; VINYL SILK - INTERNALLY					
001	One coat of emulsion paint, white; first coat to unprimed surfaces					
002	Walls over 300 mm wide:					
	plastered	0.14	2.10	0.51	**2.61**	m2
	smooth concrete	0.16	2.39	0.52	**2.91**	m2
	fibre cement	0.16	2.39	0.62	**3.01**	m2
	embossed or textured papered	0.17	2.54	0.62	**3.16**	m2
	cement rendered	0.14	2.10	0.62	**2.72**	m2
	fair face brickwork	0.18	2.80	0.63	**3.43**	m2
	fair face blockwork	0.22	3.35	0.72	**4.07**	m2
003	Walls 3.50 - 5.00 m high where ceiling is of dissimilar finish; over 300 mm wide:					
	plastered	0.14	2.19	0.51	**2.70**	m2
	smooth concrete	0.17	2.51	0.52	**3.03**	m2
	fibre cement	0.17	2.51	0.62	**3.13**	m2
	embossed or textured papered	0.18	2.67	0.62	**3.29**	m2
	cement rendered	0.14	2.19	0.62	**2.81**	m2
	fair face brickwork	0.19	2.94	0.63	**3.57**	m2
	fair face blockwork	0.23	3.52	0.72	**4.24**	m2
004	Walls in staircase areas over 300 mm wide:					
	plastered	0.14	2.15	0.51	**2.66**	m2
	smooth concrete	0.16	2.47	0.52	**2.99**	m2
	fibre cement	0.16	2.47	0.62	**3.09**	m2
	embossed or textured papered	0.17	2.62	0.62	**3.24**	m2
	cement rendered	0.14	2.15	0.62	**2.77**	m2
	fair face brickwork	0.19	2.86	0.63	**3.49**	m2
	fair face blockwork	0.23	3.44	0.72	**4.16**	m2
005	Ceilings over 300 mm wide:					
	plastered	0.14	2.19	0.51	**2.70**	m2
	smooth concrete	0.17	2.51	0.52	**3.03**	m2
	fibre cement	0.17	2.51	0.62	**3.13**	m2
	embossed or textured papered	0.18	2.67	0.62	**3.29**	m2
	cement rendered	0.14	2.19	0.62	**2.81**	m2

Unit Rates	Man-Hours	Net Labour Price £	Net Mats Price £	Net Unit Price £	Unit
006 Ceilings 3.50 - 5.00 m high over 300 mm wide:					
plastered	0.15	2.31	0.51	**2.82**	m2
smooth concrete	0.17	2.65	0.52	**3.17**	m2
fibre cement	0.17	2.65	0.62	**3.27**	m2
embossed or textured papered	0.18	2.80	0.62	**3.42**	m2
cement rendered	0.15	2.31	0.62	**2.93**	m2
007 Ceilings in staircase areas over 300 mm wide:					
plastered	0.15	2.24	0.51	**2.75**	m2
smooth concrete	0.17	2.56	0.52	**3.08**	m2
fibre cement	0.17	2.56	0.62	**3.18**	m2
embossed or textured papered	0.18	2.71	0.62	**3.33**	m2
cement rendered	0.15	2.24	0.62	**2.86**	m2
008 **One coat of emulsion paint, white; second and subsequent coats**					
009 Walls over 300 mm wide:					
plastered	0.10	1.51	0.46	**1.97**	m2
smooth concrete	0.11	1.68	0.47	**2.15**	m2
fibre cement	0.11	1.68	0.57	**2.25**	m2
embossed or textured papered	0.12	1.83	0.51	**2.34**	m2
cement rendered	0.10	1.51	0.51	**2.02**	m2
fair face brickwork	0.14	2.10	0.58	**2.68**	m2
fair face blockwork	0.15	2.35	0.64	**2.99**	m2
010 Walls 3.50 - 5.00 m high where ceiling is of dissimilar finish; over 300 mm wide:					
plastered	0.11	1.60	0.46	**2.06**	m2
smooth concrete	0.12	1.75	0.47	**2.22**	m2
fibre cement	0.12	1.75	0.57	**2.32**	m2
embossed or textured papered	0.13	1.90	0.51	**2.41**	m2
cement rendered	0.11	1.60	0.51	**2.11**	m2
fair face brickwork	0.14	2.19	0.58	**2.77**	m2
fair face blockwork	0.16	2.47	0.64	**3.11**	m2
011 Walls in staircase areas over 300 mm wide:					
plastered	0.10	1.54	0.46	**2.00**	m2
smooth concrete	0.11	1.72	0.47	**2.19**	m2
fibre cement	0.11	1.72	0.57	**2.29**	m2
embossed or textured papered	0.12	1.87	0.51	**2.38**	m2
cement rendered	0.10	1.54	0.51	**2.05**	m2
fair face brickwork	0.14	2.15	0.58	**2.73**	m2
fair face blockwork	0.16	2.41	0.64	**3.05**	m2
012 Ceilings over 300 mm wide:					
plastered	0.11	1.60	0.46	**2.06**	m2
smooth concrete	0.12	1.75	0.47	**2.22**	m2
fibre cement	0.12	1.75	0.57	**2.32**	m2
embossed or textured papered	0.13	1.90	0.51	**2.41**	m2
cement rendered	0.11	1.60	0.51	**2.11**	m2

Unit Rates

		Man-Hours	Net Labour Price £	Net Mats Price £	Net Unit Price £	Unit
013	Ceilings 3.50 - 5.00 m high over 300 mm wide:					
	plastered	0.11	1.68	0.46	**2.14**	m2
	smooth concrete	0.12	1.84	0.47	**2.31**	m2
	fibre cement	0.12	1.84	0.57	**2.41**	m2
	embossed or textured papered	0.13	2.00	0.51	**2.51**	m2
	cement rendered	0.11	1.68	0.51	**2.19**	m2
014	Ceilings in staircase areas over 300 mm wide:					
	plastered	0.11	1.63	0.46	**2.09**	m2
	smooth concrete	0.12	1.81	0.47	**2.28**	m2
	fibre cement	0.12	1.81	0.57	**2.38**	m2
	embossed or textured papered	0.13	1.96	0.51	**2.47**	m2
	cement rendered	0.11	1.63	0.51	**2.14**	m2

EMULSION PAINT; VINYL MATT - INTERNALLY

015	**One coat of emulsion paint, white; first coat to unprimed surfaces**					
016	Walls over 300 mm wide:					
	plastered	0.14	2.10	0.44	**2.54**	m2
	smooth concrete	0.16	2.39	0.46	**2.85**	m2
	fibre cement	0.16	2.39	0.54	**2.93**	m2
	embossed or textured papered	0.17	2.54	0.54	**3.08**	m2
	cement rendered	0.14	2.10	0.54	**2.64**	m2
	fair face brickwork	0.18	2.80	0.55	**3.35**	m2
	fair face blockwork	0.22	3.35	0.65	**4.00**	m2
017	Walls 3.50 - 5.00 m high where ceiling is of dissimilar finish; over 300 mm wide:					
	plastered	0.14	2.19	0.44	**2.63**	m2
	smooth concrete	0.17	2.51	0.46	**2.97**	m2
	fibre cement	0.17	2.51	0.54	**3.05**	m2
	embossed or textured papered	0.18	2.67	0.54	**3.21**	m2
	cement rendered	0.14	2.19	0.54	**2.73**	m2
	fair face brickwork	0.19	2.94	0.55	**3.49**	m2
	fair face blockwork	0.23	3.52	0.65	**4.17**	m2
018	Walls in staircase areas over 300 mm wide:					
	plastered	0.14	2.15	0.44	**2.59**	m2
	smooth concrete	0.16	2.47	0.46	**2.93**	m2
	fibre cement	0.16	2.47	0.54	**3.01**	m2
	embossed or textured papered	0.17	2.62	0.54	**3.16**	m2
	cement rendered	0.14	2.15	0.54	**2.69**	m2
	fair face brickwork	0.19	2.86	0.55	**3.41**	m2
	fair face blockwork	0.23	3.44	0.65	**4.09**	m2
019	Ceilings over 300 mm wide:					
	plastered	0.14	2.19	0.44	**2.63**	m2
	smooth concrete	0.17	2.51	0.46	**2.97**	m2
	fibre cement	0.17	2.51	0.54	**3.05**	m2
	embossed or textured papered	0.18	2.67	0.54	**3.21**	m2

Unit Rates

		Man-Hours	Net Labour Price £	Net Mats Price £	Net Unit Price £	Unit
	cement rendered	0.14	2.19	0.54	**2.73**	m2
020	Ceilings 3.50 - 5.00 m high over 300 mm wide:					
	plastered	0.15	2.31	0.44	**2.75**	m2
	smooth concrete	0.17	2.65	0.46	**3.11**	m2
	fibre cement	0.17	2.65	0.54	**3.19**	m2
	embossed or textured papered	0.18	2.80	0.54	**3.34**	m2
	cement rendered	0.15	2.31	0.54	**2.85**	m2
021	Ceilings in staircase areas over 300 mm wide:					
	plastered	0.15	2.24	0.44	**2.68**	m2
	smooth concrete	0.17	2.56	0.46	**3.02**	m2
	fibre cement	0.17	2.56	0.54	**3.10**	m2
	embossed or textured papered	0.18	2.71	0.54	**3.25**	m2
	cement rendered	0.15	2.24	0.54	**2.78**	m2
022	**One coat of emulsion paint, white; second and subsequent coats**					
023	Walls over 300 mm wide:					
	plastered	0.10	1.51	0.40	**1.91**	m2
	smooth concrete	0.11	1.68	0.41	**2.09**	m2
	fibre cement	0.11	1.68	0.49	**2.17**	m2
	embossed or textured papered	0.12	1.83	0.44	**2.27**	m2
	cement rendered	0.10	1.51	0.44	**1.95**	m2
	fair face brickwork	0.14	2.10	0.50	**2.60**	m2
	fair face blockwork	0.15	2.35	0.56	**2.91**	m2
024	Walls 3.50 - 5.00 m high where ceiling is of dissimilar finish; over 300 mm wide:					
	plastered	0.11	1.60	0.40	**2.00**	m2
	smooth concrete	0.12	1.75	0.41	**2.16**	m2
	fibre cement	0.12	1.75	0.49	**2.24**	m2
	embossed or textured papered	0.13	1.90	0.44	**2.34**	m2
	cement rendered	0.11	1.60	0.44	**2.04**	m2
	fair face brickwork	0.14	2.19	0.50	**2.69**	m2
	fair face blockwork	0.16	2.47	0.56	**3.03**	m2
025	Walls in staircase areas over 300 mm wide:					
	plastered	0.10	1.54	0.40	**1.94**	m2
	smooth concrete	0.11	1.72	0.41	**2.13**	m2
	fibre cement	0.11	1.72	0.49	**2.21**	m2
	embossed or textured papered	0.12	1.87	0.44	**2.31**	m2
	cement rendered	0.10	1.54	0.44	**1.98**	m2
	fair face brickwork	0.14	2.15	0.50	**2.65**	m2
	fair face blockwork	0.16	2.41	0.56	**2.97**	m2
026	Ceilings over 300 mm wide:					
	plastered	0.11	1.60	0.40	**2.00**	m2
	smooth concrete	0.12	1.75	0.41	**2.16**	m2
	fibre cement	0.12	1.75	0.49	**2.24**	m2
	embossed or textured papered	0.13	1.90	0.44	**2.34**	m2

Unit Rates

		Man-Hours	Net Labour Price £	Net Mats Price £	Net Unit Price £	Unit
	cement rendered	0.11	1.60	0.44	**2.04**	m2
027	Ceilings 3.50 - 5.00 m high over 300 mm wide:					
	plastered	0.11	1.68	0.40	**2.08**	m2
	smooth concrete	0.12	1.84	0.41	**2.25**	m2
	fibre cement	0.12	1.84	0.49	**2.33**	m2
	embossed or textured papered	0.13	2.00	0.44	**2.44**	m2
	cement rendered	0.11	1.68	0.44	**2.12**	m2
028	Ceilings in staircase areas over 300 mm wide:					
	plastered	0.11	1.63	0.40	**2.03**	m2
	smooth concrete	0.12	1.81	0.41	**2.22**	m2
	fibre cement	0.12	1.81	0.49	**2.30**	m2
	embossed or textured papered	0.13	1.96	0.44	**2.40**	m2
	cement rendered	0.11	1.63	0.44	**2.07**	m2

ARTEX FINISH - INTERNALLY

029 One coat of Artex sealer and one coat of Artex standard compound; with stipple finish

		Man-Hours	Net Labour Price £	Net Mats Price £	Net Unit Price £	Unit
030	Walls over 300 mm wide:					
	plastered	0.33	5.03	1.33	**6.36**	m2
	smooth concrete	0.37	5.70	1.35	**7.05**	m2
	fibre cement	0.33	5.03	1.33	**6.36**	m2
	plasterboard including scrimming joints	0.36	5.53	1.45	**6.98**	m2
	cement rendered	0.37	5.70	1.44	**7.14**	m2
	fair face brickwork	0.37	5.70	1.46	**7.16**	m2
	fair face blockwork	0.37	5.70	1.67	**7.37**	m2
031	Walls 3.50 - 5.00 m high where ceiling is of dissimilar finish; over 300 mm wide:					
	plastered	0.35	5.36	1.33	**6.69**	m2
	smooth concrete	0.40	6.03	1.35	**7.38**	m2
	fibre cement	0.35	5.36	1.33	**6.69**	m2
	plasterboard including scrimming joints	0.39	5.86	1.45	**7.31**	m2
	cement rendered	0.40	6.03	1.44	**7.47**	m2
	fair face brickwork	0.40	6.03	1.46	**7.49**	m2
	fair face blockwork	0.40	6.03	1.67	**7.70**	m2
032	Walls in staircase areas over 300 mm wide:					
	plastered	0.34	5.19	1.33	**6.52**	m2
	smooth concrete	0.39	5.86	1.35	**7.21**	m2
	fibre cement	0.34	5.19	1.33	**6.52**	m2
	plasterboard including scrimming joints	0.37	5.70	1.45	**7.15**	m2
	cement rendered	0.39	5.86	1.44	**7.30**	m2
	fair face brickwork	0.39	5.86	1.46	**7.32**	m2
	fair face blockwork	0.39	5.86	1.67	**7.53**	m2
033	Ceilings over 300 mm wide:					
	plastered	0.35	5.36	1.33	**6.69**	m2
	smooth concrete	0.40	6.03	1.35	**7.38**	m2

Unit Rates

INTERNAL WORK VA

		Man-Hours	Net Labour Price £	Net Mats Price £	Net Unit Price £	Unit
	fibre cement	0.35	5.36	1.33	**6.69**	m2
	plasterboard including scrimming joints	0.39	5.86	1.45	**7.31**	m2
	cement rendered	0.40	6.03	1.44	**7.47**	m2
034	Ceilings 3.50 - 5.00 m high over 300 mm wide:					
	plastered	0.37	5.70	1.33	**7.03**	m2
	smooth concrete	0.42	6.37	1.35	**7.72**	m2
	fibre cement	0.37	5.70	1.33	**7.03**	m2
	plasterboard including scrimming joints	0.41	6.20	1.45	**7.65**	m2
	cement rendered	0.42	6.37	1.44	**7.81**	m2
035	Ceilings in staircase areas over 300 mm wide:					
	plastered	0.36	5.53	1.33	**6.86**	m2
	smooth concrete	0.41	6.20	1.35	**7.55**	m2
	fibre cement	0.36	5.53	1.33	**6.86**	m2
	plasterboard including scrimming joints	0.40	6.03	1.45	**7.48**	m2
	cement rendered	0.40	6.03	1.44	**7.47**	m2
	PRIMERS - INTERNALLY					
036	**One coat of alkali resisting primer on untreated surfaces**					
037	Walls over 300 mm wide:					
	plastered	0.18	2.80	1.40	**4.20**	m2
	smooth concrete	0.21	3.18	1.41	**4.59**	m2
	fibre cement	0.21	3.18	1.40	**4.58**	m2
	cement rendered	0.18	2.80	1.83	**4.63**	m2
	fair face brickwork	0.24	3.72	1.85	**5.57**	m2
	fair face blockwork	0.29	4.48	2.54	**7.02**	m2
038	Walls 3.50 - 5.00 m high where ceiling is of dissimilar finish; over 300 mm wide:					
	plastered	0.19	2.94	1.40	**4.34**	m2
	smooth concrete	0.22	3.35	1.41	**4.76**	m2
	fibre cement	0.22	3.35	1.40	**4.75**	m2
	cement rendered	0.19	2.94	1.83	**4.77**	m2
	fair face brickwork	0.26	3.90	1.85	**5.75**	m2
	fair face blockwork	0.31	4.69	2.54	**7.23**	m2
039	Walls in staircase areas over 300 mm wide:					
	plastered	0.19	2.86	1.40	**4.26**	m2
	smooth concrete	0.22	3.27	1.41	**4.68**	m2
	fibre cement	0.22	3.27	1.40	**4.67**	m2
	cement rendered	0.19	2.86	1.83	**4.69**	m2
	fair face brickwork	0.25	3.82	1.85	**5.67**	m2
	fair face blockwork	0.30	4.58	2.54	**7.12**	m2
040	Ceilings over 300 mm wide:					
	plastered	0.19	2.94	1.40	**4.34**	m2
	smooth concrete	0.22	3.35	1.41	**4.76**	m2
	fibre cement	0.22	3.35	1.40	**4.75**	m2

Unit Rates

		Man-Hours	Net Labour Price £	Net Mats Price £	Net Unit Price £	Unit
	cement rendered	0.19	2.94	1.83	**4.77**	m2
041	Ceilings 3.50 - 5.00 m high over 300 mm wide:					
	plastered	0.20	3.08	1.40	**4.48**	m2
	smooth concrete	0.23	3.52	1.41	**4.93**	m2
	fibre cement	0.23	3.52	1.40	**4.92**	m2
	cement rendered	0.20	3.08	1.83	**4.91**	m2
042	Ceilings in staircase areas over 300 mm wide:					
	plastered	0.20	3.00	1.40	**4.40**	m2
	smooth concrete	0.23	3.44	1.41	**4.85**	m2
	fibre cement	0.23	3.44	1.40	**4.84**	m2
	cement rendered	0.20	3.00	1.83	**4.83**	m2
043	**One coat of all-purpose primer on untreated surfaces**					
044	Walls over 300 mm wide:					
	plastered	0.18	2.80	0.35	**3.15**	m2
	smooth concrete	0.21	3.18	0.37	**3.55**	m2
	fibre cement	0.21	3.18	0.35	**3.53**	m2
	embossed or textured papered	0.22	3.34	0.46	**3.80**	m2
	cement rendered	0.18	2.80	0.46	**3.26**	m2
	fair face brickwork	0.24	3.72	0.47	**4.19**	m2
	fair face blockwork	0.29	4.48	0.65	**5.13**	m2
045	Walls 3.50 - 5.00 m high where ceiling is of dissimilar finish; over 300 mm wide:					
	plastered	0.19	2.94	0.35	**3.29**	m2
	smooth concrete	0.22	3.35	0.37	**3.72**	m2
	fibre cement	0.22	3.35	0.35	**3.70**	m2
	embossed or textured papered	0.23	3.50	0.46	**3.96**	m2
	cement rendered	0.19	2.94	0.46	**3.40**	m2
	fair face brickwork	0.26	3.90	0.47	**4.37**	m2
	fair face blockwork	0.31	4.69	0.65	**5.34**	m2
046	Walls in staircase areas over 300 mm wide:					
	plastered	0.19	2.86	0.35	**3.21**	m2
	smooth concrete	0.22	3.27	0.37	**3.64**	m2
	fibre cement	0.22	3.27	0.35	**3.62**	m2
	embossed or textured papered	0.23	3.43	0.46	**3.89**	m2
	cement rendered	0.19	2.86	0.46	**3.32**	m2
	fair face brickwork	0.25	3.82	0.47	**4.29**	m2
	fair face blockwork	0.30	4.58	0.65	**5.23**	m2
047	Ceilings over 300 mm wide:					
	plastered	0.19	2.94	0.35	**3.29**	m2
	smooth concrete	0.22	3.35	0.37	**3.72**	m2
	fibre cement	0.22	3.35	0.35	**3.70**	m2
	embossed or textured papered	0.23	3.50	0.46	**3.96**	m2
	cement rendered	0.19	2.94	0.46	**3.40**	m2

Unit Rates

		Man-Hours	Net Labour Price £	Net Mats Price £	Net Unit Price £	Unit
048	Ceilings 3.50 - 5.00 m high over 300 mm wide:					
	plastered	0.20	3.08	0.35	**3.43**	m2
	smooth concrete	0.23	3.52	0.37	**3.89**	m2
	fibre cement	0.23	3.52	0.35	**3.87**	m2
	embossed or textured papered	0.24	3.67	0.46	**4.13**	m2
	cement rendered	0.20	3.08	0.46	**3.54**	m2
049	Ceilings in staircase areas over 300 mm wide:					
	plastered	0.20	3.00	0.35	**3.35**	m2
	smooth concrete	0.23	3.44	0.37	**3.81**	m2
	fibre cement	0.23	3.44	0.35	**3.79**	m2
	embossed or textured papered	0.24	3.59	0.46	**4.05**	m2
	cement rendered	0.20	3.00	0.46	**3.46**	m2
	ALKYD BASED PAINT; UNDERCOAT - INTERNALLY					
050	**One undercoat of alkyd based paint, white; on primed surfaces**					
051	Walls over 300 mm wide:					
	plastered	0.10	1.52	0.47	**1.99**	m2
	smooth concrete	0.11	1.68	0.78	**2.46**	m2
	fibre cement	0.11	1.68	0.56	**2.24**	m2
	embossed or textured papered	0.12	1.83	0.77	**2.60**	m2
	cement rendered	0.10	1.52	0.77	**2.29**	m2
	fair face brickwork	0.14	2.10	0.90	**3.00**	m2
	fair face blockwork	0.16	2.39	0.91	**3.30**	m2
052	Walls 3.50 - 5.00 m high where ceiling is of dissimilar finish; over 300 mm wide:					
	plastered	0.11	1.61	0.47	**2.08**	m2
	smooth concrete	0.12	1.75	0.78	**2.53**	m2
	fibre cement	0.12	1.75	0.56	**2.31**	m2
	embossed or textured papered	0.13	1.90	0.77	**2.67**	m2
	cement rendered	0.11	1.61	0.77	**2.38**	m2
	fair face brickwork	0.14	2.19	1.07	**3.26**	m2
	fair face blockwork	0.17	2.51	0.91	**3.42**	m2
053	Walls in staircase areas over 300 mm wide:					
	plastered	0.10	1.55	0.47	**2.02**	m2
	smooth concrete	0.11	1.72	0.78	**2.50**	m2
	fibre cement	0.11	1.72	0.56	**2.28**	m2
	embossed or textured papered	0.12	1.87	0.77	**2.64**	m2
	cement rendered	0.10	1.55	0.77	**2.32**	m2
	fair face brickwork	0.14	2.15	0.90	**3.05**	m2
	fair face blockwork	0.16	2.47	0.91	**3.38**	m2
054	Ceilings over 300 mm wide:					
	plastered	0.11	1.61	0.47	**2.08**	m2
	smooth concrete	0.12	1.75	0.78	**2.53**	m2
	fibre cement	0.12	1.75	0.56	**2.31**	m2
	embossed or textured papered	0.13	1.90	0.77	**2.67**	m2

Unit Rates

			Man-Hours	Net Labour Price £	Net Mats Price £	Net Unit Price £	Unit
		cement rendered	0.11	1.61	0.77	**2.38**	m2
055	Ceilings 3.50 - 5.00 m high over 300 mm wide:						
		plastered	0.11	1.69	0.47	**2.16**	m2
		smooth concrete	0.12	1.84	0.78	**2.62**	m2
		fibre cement	0.12	1.84	0.56	**2.40**	m2
		embossed or textured papered	0.13	2.00	0.77	**2.77**	m2
		cement rendered	0.11	1.69	0.77	**2.46**	m2
056	Ceilings in staircase areas over 300 mm wide:						
		plastered	0.11	1.64	0.47	**2.11**	m2
		smooth concrete	0.12	1.81	0.78	**2.59**	m2
		fibre cement	0.12	1.81	0.56	**2.37**	m2
		embossed or textured papered	0.13	1.96	0.77	**2.73**	m2
		cement rendered	0.11	1.64	0.77	**2.41**	m2
	Note 057 - 070 not used.						
	ALKYD BASED PAINT; EGGSHELL FINISH - INTERNALLY						
071	**One coat of alkyd based paint, eggshell finish, first and subsequent coats**						
072	Walls over 300 mm wide:						
		plastered	0.10	1.52	0.51	**2.03**	m2
		smooth concrete	0.11	1.68	0.52	**2.20**	m2
		fibre cement	0.11	1.68	0.61	**2.29**	m2
		embossed or textured papered	0.12	1.83	0.67	**2.50**	m2
		cement rendered	0.10	1.52	0.67	**2.19**	m2
		fair face brickwork	0.10	1.52	0.75	**2.27**	m2
		fair face blockwork	0.11	1.68	0.86	**2.54**	m2
073	Walls 3.50 - 5.00 m high where ceiling is of dissimilar finish; over 300 mm wide:						
		plastered	0.11	1.61	0.51	**2.12**	m2
		smooth concrete	0.12	1.75	0.52	**2.27**	m2
		fibre cement	0.12	1.75	0.61	**2.36**	m2
		embossed or textured papered	0.13	1.90	0.67	**2.57**	m2
		cement rendered	0.11	1.61	0.67	**2.28**	m2
		fair face brickwork	0.11	1.61	0.75	**2.36**	m2
		fair face blockwork	0.12	1.75	0.86	**2.61**	m2
074	Walls in staircase areas over 300 mm wide:						
		plastered	0.10	1.55	0.51	**2.06**	m2
		smooth concrete	0.11	1.72	0.52	**2.24**	m2
		fibre cement	0.11	1.72	0.61	**2.33**	m2
		embossed or textured papered	0.12	1.87	0.67	**2.54**	m2
		cement rendered	0.10	1.55	0.67	**2.22**	m2
		fair face brickwork	0.10	1.55	0.75	**2.30**	m2
		fair face blockwork	0.11	1.72	0.86	**2.58**	m2

Unit Rates

		Man-Hours	Net Labour Price £	Net Mats Price £	Net Unit Price £	Unit
075	Ceilings over 300 mm wide:					
	plastered	0.11	1.61	0.51	2.12	m2
	smooth concrete	0.12	1.75	0.52	2.27	m2
	fibre cement	0.12	1.75	0.61	2.36	m2
	embossed or textured papered	0.13	1.90	0.67	2.57	m2
	cement rendered	0.11	1.61	0.67	2.28	m2
076	Ceilings 3.50 - 5.00 m high over 300 mm wide:					
	plastered	0.11	1.69	0.51	2.20	m2
	smooth concrete	0.12	1.84	0.52	2.36	m2
	fibre cement	0.12	1.84	0.61	2.45	m2
	embossed or textured papered	0.13	2.00	0.67	2.67	m2
	cement rendered	0.11	1.69	0.67	2.36	m2
077	Ceilings in staircase areas over 300 mm wide:					
	plastered	0.11	1.64	0.51	2.15	m2
	smooth concrete	0.12	1.81	0.52	2.33	m2
	fibre cement	0.12	1.81	0.61	2.42	m2
	embossed or textured papered	0.13	1.96	0.67	2.63	m2
	cement rendered	0.11	1.64	0.67	2.31	m2
078	**Two coats glaze finish to multicolour surfaces**					
079	Walls over 300 mm wide:					
	plastered	0.22	3.35	1.09	4.44	m2
	smooth concrete	0.22	3.35	1.11	4.46	m2
	fibre cement	0.22	3.35	1.09	4.44	m2
	cement rendered	0.23	3.50	1.42	4.92	m2
	fair face brickwork	0.28	4.26	1.58	5.84	m2
	fair face blockwork	0.30	4.57	1.79	6.36	m2
080	Walls 3.50 - 5.00 m high where ceiling is of dissimilar finish over 300 mm wide:					
	plastered	0.24	3.66	1.09	4.75	m2
	smooth concrete	0.24	3.66	1.11	4.77	m2
	fibre cement	0.24	3.66	1.09	4.75	m2
	cement rendered	0.25	3.81	1.42	5.23	m2
	fair face brickwork	0.30	4.57	1.58	6.15	m2
	fair face blockwork	0.32	4.87	1.79	6.66	m2
081	Walls in staircase areas over 300 mm wide:					
	plastered	0.24	3.66	1.09	4.75	m2
	smooth concrete	0.24	3.66	1.11	4.77	m2
	fibre cement	0.24	3.66	1.09	4.75	m2
	cement rendered	0.25	3.81	1.42	5.23	m2
	fair face brickwork	0.30	4.57	1.58	6.15	m2
	fair face blockwork	0.32	4.87	1.79	6.66	m2

Unit Rates

		Man-Hours	Net Labour Price £	Net Mats Price £	Net Unit Price £	Unit
	ALKYD BASED PAINT; GLOSS FINISH - INTERNALLY					
082	**One coat of alkyd based paint, gloss finish to undercoated surfaces**					
083	Walls over 300 mm wide:					
	plastered	0.10	1.55	0.47	**2.02**	m2
	smooth concrete	0.11	1.69	0.69	**2.38**	m2
	fibre cement	0.11	1.69	0.56	**2.25**	m2
	embossed or textured papered	0.12	1.87	0.68	**2.55**	m2
	cement rendered	0.10	1.55	0.68	**2.23**	m2
	fair face brickwork	0.11	1.69	0.78	**2.47**	m2
	fair face blockwork	0.12	1.86	0.91	**2.77**	m2
084	Walls 3.50 - 5.00 m high where ceiling is of dissimilar finish; over 300 mm wide:					
	plastered	0.11	1.64	0.47	**2.11**	m2
	smooth concrete	0.12	1.78	0.69	**2.47**	m2
	fibre cement	0.12	1.78	0.56	**2.34**	m2
	embossed or textured papered	0.13	1.93	0.68	**2.61**	m2
	cement rendered	0.11	1.64	0.68	**2.32**	m2
	fair face brickwork	0.12	1.78	0.78	**2.56**	m2
	fair face blockwork	0.13	1.96	0.91	**2.87**	m2
085	Walls in staircase areas over 300 mm wide:					
	plastered	0.11	1.60	0.47	**2.07**	m2
	smooth concrete	0.11	1.74	0.69	**2.43**	m2
	fibre cement	0.11	1.74	0.56	**2.30**	m2
	embossed or textured papered	0.12	1.89	0.68	**2.57**	m2
	cement rendered	0.11	1.60	0.68	**2.28**	m2
	fair face brickwork	0.11	1.74	0.78	**2.52**	m2
	fair face blockwork	0.13	1.90	0.91	**2.81**	m2
086	Ceilings over 300 mm wide:					
	plastered	0.11	1.64	0.47	**2.11**	m2
	smooth concrete	0.12	1.78	0.69	**2.47**	m2
	fibre cement	0.12	1.78	0.56	**2.34**	m2
	embossed or textured papered	0.13	1.93	0.68	**2.61**	m2
	cement rendered	0.11	1.64	0.68	**2.32**	m2
087	Ceilings 3.50 - 5.00 m high over 300 mm wide:					
	plastered	0.11	1.72	0.47	**2.19**	m2
	smooth concrete	0.12	1.86	0.69	**2.55**	m2
	fibre cement	0.12	1.86	0.56	**2.42**	m2
	embossed or textured papered	0.13	2.01	0.68	**2.69**	m2
	cement rendered	0.11	1.72	0.68	**2.40**	m2
088	Ceilings in staircase areas over 300 mm wide:					
	plastered	0.11	1.68	0.47	**2.15**	m2
	smooth concrete	0.12	1.83	0.69	**2.52**	m2
	fibre cement	0.12	1.83	0.56	**2.39**	m2
	embossed or textured papered	0.13	1.98	0.68	**2.66**	m2

Unit Rates

			Man-Hours	Net Labour Price £	Net Mats Price £	Net Unit Price £	Unit
		cement rendered	0.11	1.68	0.68	2.36	m2

SUNDRIES

	089	**Cutting in to line**					
	090	Flush surfaces:					
		1 Coat work	0.11	1.68	-	1.68	m
		2 Coat work	0.17	2.51	-	2.51	m
		3 Coat work	0.22	3.35	-	3.35	m

VB WALLS AND CEILINGS - INTERNALLY

ACRYLATED RUBBER PAINT; PRIMER - INTERNALLY

	001	**One coat of acrylated rubber paint; first coat as primer/undercoat; white, on untreated surfaces; applied to a dry film thickness of 55 microns**					
	002	Walls over 300 mm wide:					
		plastered	0.18	2.80	2.81	5.61	m2
		smooth concrete	0.21	3.18	2.82	6.00	m2
		fibre cement	0.21	3.18	2.81	5.99	m2
		cement rendered	0.18	2.80	3.19	5.99	m2
		fair face brickwork	0.24	3.72	3.88	7.60	m2
		fair face blockwork	0.29	4.48	4.85	9.33	m2
	003	Walls 3.50 - 5.00 m high where ceiling is of dissimilar finish; over 300 mm wide:					
		plastered	0.19	2.94	2.81	5.75	m2
		smooth concrete	0.22	3.35	2.82	6.17	m2
		fibre cement	0.22	3.35	2.81	6.16	m2
		cement rendered	0.19	2.94	3.19	6.13	m2
		fair face brickwork	0.26	3.90	3.88	7.78	m2
		fair face blockwork	0.31	4.69	4.85	9.54	m2
	004	Walls in staircase areas over 300 mm wide:					
		plastered	0.19	2.86	2.81	5.67	m2
		smooth concrete	0.22	3.27	2.82	6.09	m2
		fibre cement	0.22	3.27	2.81	6.08	m2
		cement rendered	0.19	2.86	3.19	6.05	m2
		fair face brickwork	0.25	3.82	3.88	7.70	m2
		fair face blockwork	0.30	4.58	4.85	9.43	m2
	005	Ceilings over 300 mm wide:					
		plastered	0.19	2.94	2.81	5.75	m2
		smooth concrete	0.22	3.35	2.82	6.17	m2
		fibre cement	0.22	3.35	2.81	6.16	m2
		cement rendered	0.19	2.94	3.19	6.13	m2
	006	Ceilings 3.50 - 5.00 m high over 300 mm wide:					
		plastered	0.20	3.08	2.81	5.89	m2

Unit Rates

		Man-Hours	Net Labour Price £	Net Mats Price £	Net Unit Price £	Unit
	smooth concrete	0.23	3.52	2.82	**6.34**	m2
	fibre cement	0.23	3.52	2.81	**6.33**	m2
	cement rendered	0.20	3.08	3.19	**6.27**	m2
007	Ceilings in staircase areas over 300 mm wide:					
	plastered	0.20	3.00	2.81	**5.81**	m2
	smooth concrete	0.23	3.44	2.82	**6.26**	m2
	fibre cement	0.23	3.44	2.81	**6.25**	m2
	cement rendered	0.20	3.00	3.19	**6.19**	m2
	ACRYLATED RUBBER PAINT; UNDERCOAT - INTERNALLY					
008	**One coat of acrylated rubber paint; second coat to primed surfaces; applied to a dry film thickness 100 microns**					
009	Walls over 300 mm wide:					
	plastered	0.36	5.53	9.08	**14.61**	m2
	smooth concrete	0.42	6.37	9.10	**15.47**	m2
	fibre cement	0.42	6.37	9.08	**15.45**	m2
	cement rendered	0.36	5.53	9.48	**15.01**	m2
	fair face brickwork	0.48	7.37	9.99	**17.36**	m2
	fair face blockwork	0.57	8.71	11.10	**19.81**	m2
010	Walls 3.50 - 5.00 m high where ceiling is of dissimilar finish; over 300 mm wide:					
	plastered	0.39	5.86	9.08	**14.94**	m2
	smooth concrete	0.44	6.70	9.10	**15.80**	m2
	fibre cement	0.44	6.70	9.08	**15.78**	m2
	cement rendered	0.39	5.86	9.48	**15.34**	m2
	fair face brickwork	0.51	7.71	9.99	**17.70**	m2
	fair face blockwork	0.61	9.21	11.10	**20.31**	m2
011	Walls in staircase areas over 300 mm wide:					
	plastered	0.37	5.70	9.08	**14.78**	m2
	smooth concrete	0.43	6.53	9.10	**15.63**	m2
	fibre cement	0.43	6.53	9.08	**15.61**	m2
	cement rendered	0.37	5.70	9.48	**15.18**	m2
	fair face brickwork	0.51	7.71	9.99	**17.70**	m2
	fair face blockwork	0.58	8.88	11.10	**19.98**	m2
012	Ceilings over 300 mm wide:					
	plastered	0.39	5.86	9.08	**14.94**	m2
	smooth concrete	0.44	6.70	9.10	**15.80**	m2
	fibre cement	0.44	6.70	9.08	**15.78**	m2
	cement rendered	0.39	5.86	9.48	**15.34**	m2
013	Ceilings 3.50 - 5.00 m high over 300 mm wide:					
	plastered	0.41	6.20	9.08	**15.28**	m2
	smooth concrete	0.46	7.04	9.10	**16.14**	m2
	fibre cement	0.46	7.04	9.08	**16.12**	m2
	cement rendered	0.41	6.20	9.48	**15.68**	m2

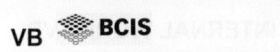

Unit Rates

		Man-Hours	Net Labour Price £	Net Mats Price £	Net Unit Price £	Unit
014	Ceilings in staircase areas over 300 mm wide:					
	plastered	0.40	6.03	9.08	**15.11**	m2
	smooth concrete	0.45	6.87	9.10	**15.97**	m2
	fibre cement	0.45	6.87	9.08	**15.95**	m2
	cement rendered	0.40	6.03	9.48	**15.51**	m2
	ACRYLATED RUBBER PAINT; FINISH - INTERNALLY					
015	**One coat of acrylated rubber paint; finish coat to coated surfaces; applied to a dry film thickness of 50 microns**					
016	Walls over 300 mm wide:					
	plastered	0.23	3.52	1.90	**5.42**	m2
	smooth concrete	0.26	4.02	2.10	**6.12**	m2
	fibre cement	0.26	4.02	1.99	**6.01**	m2
	cement rendered	0.23	3.52	2.09	**5.61**	m2
	fair face brickwork	0.31	4.69	2.57	**7.26**	m2
	fair face blockwork	0.36	5.53	2.76	**8.29**	m2
017	Walls 3.50 - 5.00 m high where ceiling is of dissimilar finish; over 300 mm wide:					
	plastered	0.24	3.69	1.90	**5.59**	m2
	smooth concrete	0.28	4.19	2.10	**6.29**	m2
	fibre cement	0.28	4.19	1.99	**6.18**	m2
	cement rendered	0.24	3.69	2.09	**5.78**	m2
	fair face brickwork	0.32	4.86	2.57	**7.43**	m2
	fair face blockwork	0.39	5.86	2.77	**8.63**	m2
018	Walls in staircase areas over 300 mm wide:					
	plastered	0.23	3.52	1.90	**5.42**	m2
	smooth concrete	0.26	4.02	2.10	**6.12**	m2
	fibre cement	0.26	4.02	1.99	**6.01**	m2
	cement rendered	0.23	3.52	2.09	**5.61**	m2
	fair face brickwork	0.32	4.86	2.57	**7.43**	m2
	fair face blockwork	0.37	5.70	2.77	**8.47**	m2
019	Ceilings over 300 mm wide:					
	plastered	0.24	3.69	1.90	**5.59**	m2
	smooth concrete	0.28	4.19	2.10	**6.29**	m2
	fibre cement	0.28	4.19	1.99	**6.18**	m2
	cement rendered	0.24	3.69	2.09	**5.78**	m2
020	Ceilings 3.50 - 5.00 m high over 300 mm wide:					
	plastered	0.25	3.85	1.90	**5.75**	m2
	smooth concrete	0.29	4.36	2.10	**6.46**	m2
	fibre cement	0.29	4.36	1.99	**6.35**	m2
	cement rendered	0.25	3.85	2.09	**5.94**	m2
021	Ceilings in staircase areas over 300 mm wide:					
	plastered	0.24	3.69	1.90	**5.59**	m2
	smooth concrete	0.29	4.36	2.10	**6.46**	m2

Unit Rates

	Man-Hours	Net Labour Price £	Net Mats Price £	Net Unit Price £	Unit
fibre cement	0.29	4.36	1.99	**6.35**	m2
cement rendered	0.24	3.69	2.09	**5.78**	m2

ANTI-BACTERIAL PAINT SYSTEM - INTERNALLY

022 **Prepare; apply one coat primer/adhesive high performance water-based, two component epoxy coating**

023 Walls over 300 mm wide:

	Man-Hours	Net Labour Price £	Net Mats Price £	Net Unit Price £	Unit
plastered	0.12	1.75	1.97	**3.72**	m2
smooth concrete	0.13	1.93	1.97	**3.90**	m2
fibre cement	0.14	2.13	2.09	**4.22**	m2
embossed or textured paper	0.14	2.13	2.09	**4.22**	m2
cement render	0.15	2.35	2.09	**4.44**	m2
fair face brickwork	0.17	2.57	2.09	**4.66**	m2
fair face blockwork	0.19	2.83	2.09	**4.92**	m2

024 Walls 3.50 - 5.00 m high where ceiling is of dissimilar finish; over 300 mm wide:

	Man-Hours	Net Labour Price £	Net Mats Price £	Net Unit Price £	Unit
plastered	0.12	1.89	2.24	**4.13**	m2
smooth concrete	0.14	2.09	2.09	**4.18**	m2
fibre cement	0.15	2.30	2.09	**4.39**	m2
embossed or textured paper	0.16	2.36	2.09	**4.45**	m2
cement render	0.17	2.53	2.09	**4.62**	m2
fair face brickwork	0.18	2.77	2.09	**4.86**	m2
fair face blockwork	0.20	3.05	2.24	**5.29**	m2

025 Walls in staircase area over 300 mm wide:

	Man-Hours	Net Labour Price £	Net Mats Price £	Net Unit Price £	Unit
plastered	0.13	2.03	1.97	**4.00**	m2
smooth concrete	0.15	2.24	1.97	**4.21**	m2
fibre cement	0.16	2.47	2.09	**4.56**	m2
embossed or textured paper	0.03	0.46	2.09	**2.55**	m2
cement render	0.18	2.71	2.09	**4.80**	m2
fair face brickwork	0.20	2.99	2.09	**5.08**	m2
fair face blockwork	0.22	3.27	2.24	**5.51**	m2

026 Ceilings over 300 mm wide:

	Man-Hours	Net Labour Price £	Net Mats Price £	Net Unit Price £	Unit
plastered	0.13	1.93	1.97	**3.90**	m2
smooth concrete	0.14	2.13	1.97	**4.10**	m2
fibre cement	0.15	2.35	2.09	**4.44**	m2
cement render	0.17	2.57	2.09	**4.66**	m2
fair face brickwork	0.19	2.83	2.09	**4.92**	m2
fair face blockwork	0.21	3.12	2.24	**5.36**	m2

027 Ceilings 3.50- 5.00 m high over 300 mm wide:

	Man-Hours	Net Labour Price £	Net Mats Price £	Net Unit Price £	Unit
plastered	0.13	2.03	1.97	**4.00**	m2
smooth concrete	0.15	2.24	1.97	**4.21**	m2
fibre cement	0.16	2.47	2.09	**4.56**	m2
cement render	0.18	2.70	2.09	**4.79**	m2
fair face brickwork	0.20	2.97	2.09	**5.06**	m2
fair face blockwork	0.22	3.27	2.24	**5.51**	m2

Unit Rates

		Man-Hours	Net Labour Price £	Net Mats Price £	Net Unit Price £	Unit
028	Ceilings in staircase areas over 300 mm wide:					
	plastered	0.13	2.03	1.97	**4.00**	m2
	smooth concrete	0.15	2.24	1.97	**4.21**	m2
	fibre cement	0.16	2.47	2.04	**4.51**	m2
	cement render	0.18	2.70	2.04	**4.74**	m2
	fair face brickwork	0.20	2.97	2.04	**5.01**	m2
	fair face blockwork	0.22	3.27	2.24	**5.51**	m2
029	**One coat water based hygiene paint, two pack polyurethane system, to primed surface; and subsequent coats**					
030	Walls over 300 mm wide:					
	plastered	0.10	1.52	2.51	**4.03**	m2
	smooth concrete	0.11	1.68	2.51	**4.19**	m2
	fibre cement	0.12	1.84	2.67	**4.51**	m2
	embossed or textured paper	0.12	1.84	2.67	**4.51**	m2
	cement render	0.13	2.03	2.67	**4.70**	m2
	fair face brickwork	0.15	2.22	2.67	**4.89**	m2
	fair face blockwork	0.16	2.45	2.83	**5.28**	m2
031	Walls 3.50 - 5.00 m high where ceiling is of dissimilar finish; over 300 mm wide:					
	plastered	0.11	1.64	2.51	**4.15**	m2
	smooth concrete	0.12	1.80	2.51	**4.31**	m2
	fibre cement	0.13	1.98	2.67	**4.65**	m2
	embossed or textured paper	0.13	1.98	2.67	**4.65**	m2
	cement render	0.14	2.18	2.67	**4.85**	m2
	fair face brickwork	0.16	2.39	2.67	**5.06**	m2
	fair face blockwork	0.17	2.63	2.83	**5.46**	m2
032	Walls in staircase area over 300 mm wide:					
	plastered	0.12	1.77	2.51	**4.28**	m2
	smooth concrete	0.13	1.93	2.51	**4.44**	m2
	fibre cement	0.14	2.13	2.67	**4.80**	m2
	embossed or textured paper	0.14	2.13	2.67	**4.80**	m2
	cement render	0.15	2.35	2.67	**5.02**	m2
	fair face brickwork	0.17	2.57	2.67	**5.24**	m2
	fair face blockwork	0.19	2.83	2.83	**5.66**	m2
033	Ceilings over 300 mm wide:					
	plastered	0.11	1.68	2.51	**4.19**	m2
	smooth concrete	0.12	1.84	2.51	**4.35**	m2
	fibre cement	0.13	2.03	2.67	**4.70**	m2
	cement render	0.15	2.22	2.67	**4.89**	m2
	fair face brickwork	0.16	2.45	2.67	**5.12**	m2
	fair face blockwork	0.18	2.70	2.83	**5.53**	m2
034	Ceilings 3.50 - 5.00 m high over 300 mm wide:					
	plastered	0.11	1.68	2.51	**4.19**	m2
	smooth concrete	0.12	1.84	2.51	**4.35**	m2
	fibre cement	0.13	2.03	2.67	**4.70**	m2

Unit Rates

		Man-Hours	Net Labour Price £	Net Mats Price £	Net Unit Price £	Unit
	cement render	0.15	2.22	2.67	**4.89**	m2
	fair face brickwork	0.16	2.45	2.67	**5.12**	m2
	fair face blockwork	0.18	2.70	2.83	**5.53**	m2
035	Ceilings in staircase areas over 300 mm wide:					
	plastered	0.13	1.95	2.51	**4.46**	m2
	smooth concrete	0.14	2.13	2.51	**4.64**	m2
	fibre cement	0.15	2.35	2.67	**5.02**	m2
	cement render	0.17	2.57	2.67	**5.24**	m2
	fair face brickwork	0.19	2.83	2.67	**5.50**	m2
	fair face blockwork	0.21	3.12	2.83	**5.95**	m2

INTUMESCENT PAINT - INTERNALLY

036	**Prepare; one coat Thermoguard Wallcoat to painted**					
037	Walls over 300 mm wide:					
	plastered	0.13	1.98	2.22	**4.20**	m2
	smooth concrete	0.14	2.18	2.50	**4.68**	m2
	fibre cement	0.16	2.39	2.50	**4.89**	m2
	embossed or textured paper	0.12	1.84	2.50	**4.34**	m2
	cement render	0.17	2.63	2.50	**5.13**	m2
	fair face brickwork	0.19	2.89	2.50	**5.39**	m2
	fair face blockwork	0.21	3.18	2.85	**6.03**	m2
038	Walls 3.50 - 5.00 m high where ceiling is of dissimilar finish; over 300 mm wide:					
	plastered	0.14	2.09	2.22	**4.31**	m2
	smooth concrete	0.15	2.35	2.50	**4.85**	m2
	fibre cement	0.17	2.57	2.50	**5.07**	m2
	embossed or textured paper	0.13	1.98	2.50	**4.48**	m2
	cement render	0.19	2.83	2.50	**5.33**	m2
	fair face brickwork	0.20	3.11	2.50	**5.61**	m2
	fair face blockwork	0.23	3.43	2.50	**5.93**	m2
039	Walls in staircase area over 300 mm wide:					
	plastered	0.15	2.24	2.22	**4.46**	m2
	smooth concrete	0.17	2.53	2.50	**5.03**	m2
	fibre cement	0.18	2.77	2.50	**5.27**	m2
	embossed or textured paper	0.14	2.13	2.50	**4.63**	m2
	cement render	0.20	3.05	2.50	**5.55**	m2
	fair face brickwork	0.22	3.34	2.50	**5.84**	m2
	fair face blockwork	0.24	3.69	2.50	**6.19**	m2
040	Ceilings over 300 mm wide:					
	plastered	0.14	2.18	2.50	**4.68**	m2
	smooth concrete	0.16	2.39	2.50	**4.89**	m2
	fibre cement	0.17	2.63	2.50	**5.13**	m2
	cement render	0.19	2.89	2.50	**5.39**	m2
	fair face brickwork	0.21	3.18	2.50	**5.68**	m2
	fair face blockwork	0.23	3.50	2.50	**6.00**	m2

Unit Rates

		Man-Hours	Net Labour Price £	Net Mats Price £	Net Unit Price £	Unit
041	Ceilings 3.50 - 5.00 m high over 300 mm wide:					
	plastered	0.15	2.28	2.22	**4.50**	m2
	smooth concrete	0.17	2.51	2.50	**5.01**	m2
	fibre cement	0.18	2.77	2.50	**5.27**	m2
	cement render	0.20	3.05	2.50	**5.55**	m2
	fair face brickwork	0.22	3.34	2.50	**5.84**	m2
	fair face blockwork	0.24	3.69	2.50	**6.19**	m2
042	Ceilings in staircase areas over 300 mm wide:					
	plastered	0.16	2.41	2.50	**4.91**	m2
	smooth concrete	0.17	2.63	2.50	**5.13**	m2
	fibre cement	0.19	2.91	2.50	**5.41**	m2
	cement render	0.21	3.20	2.50	**5.70**	m2
	fair face brickwork	0.23	3.50	2.50	**6.00**	m2
	fair face blockwork	0.25	3.87	2.50	**6.37**	m2
043	**Prepare; second and subsequent coat 'Thermoguard flame retardant acrylic', matt/eggshell**					
044	Walls over 300 mm wide:					
	plastered	0.11	1.68	1.57	**3.25**	m2
	smooth concrete	0.12	1.84	1.57	**3.41**	m2
	fibre cement	0.13	2.03	1.74	**3.77**	m2
	embossed or textured paper	0.12	1.84	1.74	**3.58**	m2
	cement render	0.15	2.22	1.74	**3.96**	m2
	fair face brickwork	0.16	2.45	1.74	**4.19**	m2
	fair face blockwork	0.18	2.70	1.95	**4.65**	m2
045	Walls 3.50 - 5.00 m high where ceiling is of dissimilar finish; over 300 mm wide:					
	plastered	0.12	1.77	1.57	**3.34**	m2
	smooth concrete	0.13	1.98	1.57	**3.55**	m2
	fibre cement	0.14	2.18	1.74	**3.92**	m2
	embossed or textured paper	0.13	1.98	1.74	**3.72**	m2
	cement render	0.16	2.39	1.74	**4.13**	m2
	fair face brickwork	0.17	2.63	1.74	**4.37**	m2
	fair face blockwork	0.19	2.89	1.95	**4.84**	m2
046	Walls in staircase area over 300 mm wide:					
	plastered	0.13	1.90	1.57	**3.47**	m2
	smooth concrete	0.14	2.13	1.57	**3.70**	m2
	fibre cement	0.15	2.35	1.74	**4.09**	m2
	embossed or textured paper	0.14	2.13	1.74	**3.87**	m2
	cement render	0.17	2.57	1.74	**4.31**	m2
	fair face brickwork	0.19	2.83	1.74	**4.57**	m2
	fair face blockwork	0.20	3.11	1.95	**5.06**	m2
047	Ceilings over 300 mm wide:					
	plastered	0.12	1.84	1.57	**3.41**	m2
	smooth concrete	0.13	2.03	1.57	**3.60**	m2
	fibre cement	0.15	2.22	1.74	**3.96**	m2
	cement render	0.16	2.45	1.74	**4.19**	m2

VC	Unit Rates	Man-Hours	Net Labour Price £	Net Mats Price £	Net Unit Price £	Unit
	fair face brickwork	0.18	2.70	1.74	**4.44**	m2
	fair face blockwork	0.20	2.97	1.95	**4.92**	m2
048	Ceilings 3.50 - 5.00 m high over 300 mm wide:					
	plastered	0.13	1.93	1.57	**3.50**	m2
	smooth concrete	0.14	2.13	1.57	**3.70**	m2
	fibre cement	0.15	2.33	1.74	**4.07**	m2
	cement render	0.17	2.57	1.74	**4.31**	m2
	fair face brickwork	0.19	2.83	1.74	**4.57**	m2
	fair face blockwork	0.21	3.12	1.95	**5.07**	m2
049	Ceilings in staircase areas over 300 mm wide:					
	plastered	0.13	2.03	1.57	**3.60**	m2
	smooth concrete	0.15	2.24	1.57	**3.81**	m2
	fibre cement	0.16	2.45	1.74	**4.19**	m2
	cement render	0.18	2.70	1.74	**4.44**	m2
	fair face brickwork	0.20	2.97	1.74	**4.71**	m2
	fair face blockwork	0.22	3.27	1.95	**5.22**	m2
VC	**WALLS AND CEILINGS, SPRAYED FINISHES - INTERNALLY**					
	EMULSION PAINT; VINYL SILK - SPRAYED INTERNALLY					
001	**One coat of emulsion paint, white; first coat to unprimed surfaces**					
002	Walls over 300 mm wide:					
	plastered	0.05	1.14	0.66	**1.80**	m2
	smooth concrete	0.05	1.26	0.66	**1.92**	m2
	fibre cement	0.05	1.14	0.80	**1.94**	m2
	embossed or textured papered	0.05	1.14	0.80	**1.94**	m2
	cement rendered	0.06	1.43	0.66	**2.09**	m2
	fair face brickwork	0.07	1.60	0.90	**2.50**	m2
	fair face blockwork	0.07	1.69	0.90	**2.59**	m2
003	Walls 3.50 - 5.00 m high where ceiling is of dissimilar finish; over 300 mm wide:					
	plastered	0.05	1.19	0.66	**1.85**	m2
	smooth concrete	0.06	1.31	0.66	**1.97**	m2
	fibre cement	0.05	1.19	0.80	**1.99**	m2
	embossed or textured papered	0.05	1.19	0.80	**1.99**	m2
	cement rendered	0.06	1.53	0.66	**2.19**	m2
	fair face brickwork	0.07	1.69	0.90	**2.59**	m2
	fair face blockwork	0.07	1.76	0.90	**2.66**	m2
004	Walls in staircase areas over 300 mm wide:					
	plastered	0.05	1.17	0.66	**1.83**	m2
	smooth concrete	0.05	1.29	0.66	**1.95**	m2
	fibre cement	0.05	1.17	0.80	**1.97**	m2
	embossed or textured papered	0.05	1.17	0.80	**1.97**	m2
	cement rendered	0.06	1.45	0.66	**2.11**	m2
	fair face brickwork	0.07	1.62	0.90	**2.52**	m2

Unit Rates

INTERNAL WORK VC

		Man-Hours	Net Labour Price £	Net Mats Price £	Net Unit Price £	Unit
	fair face blockwork	0.07	1.72	0.90	**2.62**	m2
005	Ceilings over 300 mm wide:					
	plastered	0.05	1.19	0.66	**1.85**	m2
	smooth concrete	0.06	1.31	0.66	**1.97**	m2
	fibre cement	0.05	1.19	0.80	**1.99**	m2
	embossed or textured papered	0.05	1.19	0.80	**1.99**	m2
	cement rendered	0.06	1.53	0.66	**2.19**	m2
006	Ceilings 3.50 - 5.00 m high over 300 mm wide:					
	plastered	0.05	1.26	0.66	**1.92**	m2
	smooth concrete	0.06	1.38	0.66	**2.04**	m2
	fibre cement	0.05	1.26	0.80	**2.06**	m2
	embossed or textured papered	0.05	1.26	0.80	**2.06**	m2
	cement rendered	0.07	1.60	0.66	**2.26**	m2
007	Ceilings in staircase areas over 300 mm wide:					
	plastered	0.05	1.24	0.66	**1.90**	m2
	smooth concrete	0.06	1.34	0.66	**2.00**	m2
	fibre cement	0.05	1.24	0.80	**2.04**	m2
	embossed or textured papered	0.05	1.24	0.80	**2.04**	m2
	cement rendered	0.07	1.55	0.66	**2.21**	m2
008	**One coat of emulsion paint, white; second and subsequent coats**					
009	Walls over 300 mm wide:					
	plastered	0.04	1.03	0.62	**1.65**	m2
	smooth concrete	0.05	1.14	0.62	**1.76**	m2
	fibre cement	0.04	1.03	0.76	**1.79**	m2
	embossed or textured papered	0.04	1.03	0.62	**1.65**	m2
	cement rendered	0.05	1.29	0.76	**2.05**	m2
	fair face brickwork	0.06	1.48	0.85	**2.33**	m2
	fair face blockwork	0.06	1.53	0.85	**2.38**	m2
010	Walls 3.50 - 5.00 m high where ceiling is of dissimilar finish; over 300 mm wide:					
	plastered	0.05	1.10	0.62	**1.72**	m2
	smooth concrete	0.05	1.19	0.62	**1.81**	m2
	fibre cement	0.05	1.10	0.76	**1.86**	m2
	embossed or textured papered	0.05	1.10	0.62	**1.72**	m2
	cement rendered	0.06	1.34	0.76	**2.10**	m2
	fair face brickwork	0.07	1.57	0.85	**2.42**	m2
	fair face blockwork	0.07	1.60	0.85	**2.45**	m2
011	Walls in staircase areas over 300 mm wide:					
	plastered	0.04	1.05	0.62	**1.67**	m2
	smooth concrete	0.05	1.17	0.62	**1.79**	m2
	fibre cement	0.04	1.05	0.76	**1.81**	m2
	embossed or textured papered	0.04	1.05	0.62	**1.67**	m2
	cement rendered	0.06	1.31	0.76	**2.07**	m2
	fair face brickwork	0.06	1.53	0.85	**2.38**	m2

		Man-Hours	Net Labour Price £	Net Mats Price £	Net Unit Price £	Unit
	fair face blockwork	0.07	1.55	0.85	**2.40**	m2
012	Ceilings over 300 mm wide:					
	plastered	0.05	1.10	0.64	**1.74**	m2
	smooth concrete	0.05	1.19	0.62	**1.81**	m2
	fibre cement	0.05	1.10	0.76	**1.86**	m2
	embossed or textured papered	0.05	1.10	0.62	**1.72**	m2
	cement rendered	0.06	1.34	0.76	**2.10**	m2
013	Ceilings 3.50 - 5.00 m high over 300 mm wide:					
	plastered	0.05	1.14	0.62	**1.76**	m2
	smooth concrete	0.05	1.26	0.62	**1.88**	m2
	fibre cement	0.05	1.14	0.76	**1.90**	m2
	embossed or textured papered	0.05	1.14	0.62	**1.76**	m2
	cement rendered	0.06	1.41	0.76	**2.17**	m2
014	Ceilings in staircase areas over 300 mm wide:					
	plastered	0.05	1.12	0.62	**1.74**	m2
	smooth concrete	0.05	1.24	0.62	**1.86**	m2
	fibre cement	0.05	1.12	0.76	**1.88**	m2
	embossed or textured papered	0.05	1.12	0.62	**1.74**	m2
	cement rendered	0.06	1.38	0.76	**2.14**	m2
	EMULSION PAINT; VINYL MATT - SPRAYED INTERNALLY					
015	**One coat of emulsion paint, white; first coat to unprimed surface**					
016	Walls over 300 mm wide:					
	plastered	0.05	1.14	0.57	**1.71**	m2
	smooth concrete	0.05	1.26	0.57	**1.83**	m2
	fibre cement	0.05	1.14	0.69	**1.83**	m2
	embossed or textured papered	0.05	1.14	0.69	**1.83**	m2
	cement rendered	0.06	1.43	0.57	**2.00**	m2
	fair face brickwork	0.07	1.60	0.78	**2.38**	m2
	fair face blockwork	0.07	1.69	0.78	**2.47**	m2
017	Walls 3.50 - 5.00 m high where ceiling is of dissimilar finish; over 300 mm wide:					
	plastered	0.05	1.19	0.57	**1.76**	m2
	smooth concrete	0.06	1.31	0.57	**1.88**	m2
	fibre cement	0.05	1.19	0.69	**1.88**	m2
	embossed or textured papered	0.05	1.19	0.69	**1.88**	m2
	cement rendered	0.06	1.53	0.57	**2.10**	m2
	fair face brickwork	0.07	1.69	0.78	**2.47**	m2
	fair face blockwork	0.07	1.76	0.78	**2.54**	m2
018	Walls in staircase areas over 300 mm wide:					
	plastered	0.05	1.17	0.57	**1.74**	m2
	smooth concrete	0.05	1.29	0.57	**1.86**	m2
	fibre cement	0.05	1.17	0.69	**1.86**	m2
	embossed or textured papered	0.05	1.17	0.69	**1.86**	m2

Unit Rates

		Man-Hours	Net Labour Price £	Net Mats Price £	Net Unit Price £	Unit
	cement rendered	0.06	1.45	0.57	**2.02**	m2
	fair face brickwork	0.07	1.62	0.78	**2.40**	m2
	fair face blockwork	0.07	1.72	0.78	**2.50**	m2
019	Ceilings over 300 mm wide:					
	plastered	0.05	1.19	0.57	**1.76**	m2
	smooth concrete	0.06	1.31	0.57	**1.88**	m2
	fibre cement	0.05	1.19	0.69	**1.88**	m2
	embossed or textured papered	0.05	1.19	0.69	**1.88**	m2
	cement rendered	0.06	1.53	0.57	**2.10**	m2
020	Ceilings 3.50 - 5.00 m high over 300 mm wide:					
	plastered	0.05	1.26	0.57	**1.83**	m2
	smooth concrete	0.06	1.38	0.57	**1.95**	m2
	fibre cement	0.05	1.26	0.69	**1.95**	m2
	embossed or textured papered	0.05	1.26	0.70	**1.96**	m2
	cement rendered	0.07	1.60	0.57	**2.17**	m2
021	Ceilings in staircase areas over 300 mm wide:					
	plastered	0.05	1.24	0.57	**1.81**	m2
	smooth concrete	0.06	1.34	0.57	**1.91**	m2
	fibre cement	0.05	1.24	0.69	**1.93**	m2
	embossed or textured papered	0.05	1.24	0.69	**1.93**	m2
	cement rendered	0.07	1.55	0.57	**2.12**	m2
022	**One coat of emulsion paint, white; second and subsequent coats**					
023	Walls over 300 mm wide:					
	plastered	0.04	1.03	0.53	**1.56**	m2
	smooth concrete	0.05	1.14	0.53	**1.67**	m2
	fibre cement	0.04	1.03	0.65	**1.68**	m2
	embossed or textured papered	0.04	1.03	0.53	**1.56**	m2
	cement rendered	0.05	1.29	0.65	**1.94**	m2
	fair face brickwork	0.06	1.48	0.74	**2.22**	m2
	fair face blockwork	0.06	1.53	0.74	**2.27**	m2
024	Walls 3.50 - 5.00 m high where ceiling is of dissimilar finish; over 300 mm wide:					
	plastered	0.05	1.10	0.53	**1.63**	m2
	smooth concrete	0.05	1.19	0.53	**1.72**	m2
	fibre cement	0.05	1.10	0.65	**1.75**	m2
	embossed or textured papered	0.05	1.10	0.53	**1.63**	m2
	cement rendered	0.06	1.34	0.65	**1.99**	m2
	fair face brickwork	0.07	1.57	0.74	**2.31**	m2
	fair face blockwork	0.07	1.60	0.74	**2.34**	m2
025	Walls in staircase areas over 300 mm wide:					
	plastered	0.04	1.05	0.53	**1.58**	m2
	smooth concrete	0.05	1.17	0.53	**1.70**	m2
	fibre cement	0.04	1.05	0.65	**1.70**	m2
	embossed or textured papered	0.04	1.05	0.53	**1.58**	m2

Unit Rates

		Man-Hours	Net Labour Price £	Net Mats Price £	Net Unit Price £	Unit
	cement rendered	0.06	1.31	0.65	**1.96**	m2
	fair face brickwork	0.06	1.53	0.74	**2.27**	m2
	fair face blockwork	0.07	1.55	0.74	**2.29**	m2
026	Ceilings over 300 mm wide:					
	plastered	0.05	1.10	0.53	**1.63**	m2
	smooth concrete	0.05	1.19	0.53	**1.72**	m2
	fibre cement	0.05	1.10	0.65	**1.75**	m2
	embossed or textured papered	0.05	1.10	0.53	**1.63**	m2
	cement rendered	0.06	1.43	0.65	**2.08**	m2
027	Ceilings 3.50 - 5.00 m high over 300 mm wide:					
	plastered	0.05	1.14	0.53	**1.67**	m2
	smooth concrete	0.05	1.26	0.53	**1.79**	m2
	fibre cement	0.05	1.14	0.65	**1.79**	m2
	embossed or textured papered	0.05	1.14	0.53	**1.67**	m2
	cement rendered	0.06	1.41	0.65	**2.06**	m2
028	Ceilings in staircase areas over 300 mm wide:					
	plastered	0.05	1.12	0.53	**1.65**	m2
	smooth concrete	0.05	1.24	0.53	**1.77**	m2
	fibre cement	0.05	1.12	0.65	**1.77**	m2
	embossed or textured papered	0.05	1.12	0.53	**1.65**	m2
	cement rendered	0.06	1.38	0.65	**2.03**	m2
	Note 029 - 042 not used					
	ALKYD BASED PAINT; UNDERCOAT - SPRAYED INTERNALLY					
043	**One undercoat of alkyd based paint, white; on primed surfaces**					
044	Walls over 300 mm wide:					
	plastered	0.05	1.14	0.60	**1.74**	m2
	smooth concrete	0.05	1.26	0.99	**2.25**	m2
	fibre cement	0.05	1.14	0.71	**1.85**	m2
	embossed or textured papered	0.05	1.14	0.60	**1.74**	m2
	cement rendered	0.06	1.43	0.99	**2.42**	m2
	fair face brickwork	0.07	1.60	1.15	**2.75**	m2
	fair face blockwork	0.07	1.69	1.15	**2.84**	m2
045	Walls 3.50 - 5.00 m high where ceiling is of dissimilar finish; over 300 mm wide:					
	plastered	0.05	1.19	0.60	**1.79**	m2
	smooth concrete	0.06	1.31	0.99	**2.30**	m2
	fibre cement	0.05	1.19	0.71	**1.90**	m2
	embossed or textured papered	0.05	1.19	0.60	**1.79**	m2
	cement rendered	0.06	1.53	0.99	**2.52**	m2
	fair face brickwork	0.07	1.69	1.15	**2.84**	m2
	fair face blockwork	0.07	1.76	1.15	**2.91**	m2

		Man-Hours	Net Labour Price £	Net Mats Price £	Net Unit Price £	Unit
046	Walls in staircase areas over 300 mm wide:					
	plastered	0.05	1.17	0.60	**1.77**	m2
	smooth concrete	0.05	1.29	0.99	**2.28**	m2
	fibre cement	0.05	1.17	0.71	**1.88**	m2
	embossed or textured papered	0.05	1.17	0.60	**1.77**	m2
	cement rendered	0.06	1.45	0.99	**2.44**	m2
	fair face brickwork	0.07	1.62	1.15	**2.77**	m2
	fair face blockwork	0.07	1.72	1.15	**2.87**	m2
047	Ceilings over 300 mm wide:					
	plastered	0.05	1.19	0.60	**1.79**	m2
	smooth concrete	0.06	1.31	0.99	**2.30**	m2
	fibre cement	0.05	1.19	0.71	**1.90**	m2
	embossed or textured papered	0.05	1.19	0.60	**1.79**	m2
	cement rendered	0.05	1.19	0.99	**2.18**	m2
048	Ceilings 3.50 - 5.00 mm high over 300 mm wide:					
	plastered	0.05	1.26	0.60	**1.86**	m2
	smooth concrete	0.06	1.38	0.99	**2.37**	m2
	fibre cement	0.05	1.26	0.71	**1.97**	m2
	embossed or textured papered	0.05	1.26	0.60	**1.86**	m2
	cement rendered	0.07	1.60	0.99	**2.59**	m2
049	Ceilings in staircase areas over 300 mm wide:					
	plastered	0.05	1.24	0.60	**1.84**	m2
	smooth concrete	0.06	1.34	0.99	**2.33**	m2
	fibre cement	0.05	1.24	0.71	**1.95**	m2
	embossed or textured papered	0.05	1.24	0.60	**1.84**	m2
	cement rendered	0.07	1.55	0.99	**2.54**	m2

ALKYD BASED PAINT; EGGSHELL FINISH - SPRAYED

		Man-Hours	Net Labour Price £	Net Mats Price £	Net Unit Price £	Unit
050	**One coat of alkyd based paint, eggshell finish, first and subsequent coats**					
051	Walls over 300 mm wide:					
	plastered	0.05	1.19	0.61	**1.80**	m2
	smooth concrete	0.06	1.31	0.62	**1.93**	m2
	fibre cement	0.05	1.19	0.73	**1.92**	m2
	embossed or textured papered	0.05	1.19	0.61	**1.80**	m2
	cement rendered	0.06	1.53	0.81	**2.34**	m2
	fair face brickwork	0.07	1.67	0.90	**2.57**	m2
	fair face blockwork	0.07	1.76	1.02	**2.78**	m2
052	Walls 3.50 - 5.00 m high where ceiling is of dissimilar finish; over 300 mm wide:					
	plastered	0.05	1.26	0.61	**1.87**	m2
	smooth concrete	0.06	1.38	0.61	**1.99**	m2
	fibre cement	0.05	1.26	0.73	**1.99**	m2
	embossed or textured papered	0.05	1.26	0.61	**1.87**	m2
	cement rendered	0.07	1.60	0.81	**2.41**	m2
	fair face brickwork	0.07	1.74	0.90	**2.64**	m2

Unit Rates

		Man-Hours	Net Labour Price £	Net Mats Price £	Net Unit Price £	Unit
	fair face blockwork	0.08	1.86	1.02	**2.88**	m2
053	Walls in staircase areas over 300 mm wide:					
	plastered	0.05	1.24	0.61	**1.85**	m2
	smooth concrete	0.06	1.34	0.61	**1.95**	m2
	fibre cement	0.05	1.24	0.73	**1.97**	m2
	embossed or textured papered	0.05	1.24	0.61	**1.85**	m2
	cement rendered	0.07	1.55	0.81	**2.36**	m2
	fair face brickwork	0.07	1.69	0.90	**2.59**	m2
	fair face blockwork	0.08	1.84	1.02	**2.86**	m2
054	Ceilings over 300 mm wide:					
	plastered	0.05	1.26	0.61	**1.87**	m2
	smooth concrete	0.06	1.38	0.61	**1.99**	m2
	fibre cement	0.05	1.26	0.73	**1.99**	m2
	embossed or textured papered	0.05	1.26	0.61	**1.87**	m2
	cement rendered	0.07	1.60	0.81	**2.41**	m2
055	Ceilings 3.50 - 5.00 m high over 300 mm wide:					
	plastered	0.06	1.31	0.61	**1.92**	m2
	smooth concrete	0.06	1.43	0.61	**2.04**	m2
	fibre cement	0.06	1.31	0.73	**2.04**	m2
	embossed or textured papered	0.06	1.31	0.61	**1.92**	m2
	cement rendered	0.07	1.69	0.81	**2.50**	m2
056	Ceilings in staircase areas over 300 mm wide:					
	plastered	0.05	1.29	0.61	**1.90**	m2
	smooth concrete	0.06	1.41	0.61	**2.02**	m2
	fibre cement	0.05	1.29	0.73	**2.02**	m2
	embossed or textured papered	0.05	1.29	0.61	**1.90**	m2
	cement rendered	0.07	1.62	0.81	**2.43**	m2
	ALKYD BASED PAINT; GLOSS FINISH - SPRAYED INTERNALLY					
057	**One coat of alkyd based paint gloss finish to undercoated surfaces**					
058	Walls over 300 mm wide:					
	plastered	0.05	1.19	0.56	**1.75**	m2
	smooth concrete	0.06	1.31	0.82	**2.13**	m2
	fibre cement	0.05	1.19	0.67	**1.86**	m2
	embossed or textured papered	0.05	1.19	0.56	**1.75**	m2
	cement rendered	0.06	1.53	0.82	**2.35**	m2
	fair face brickwork	0.07	1.67	0.93	**2.60**	m2
	fair face blockwork	0.07	1.76	1.08	**2.84**	m2
059	Walls 3.50 - 5.00 m high where ceiling is of dissimilar finish; over 300 mm wide:					
	plastered	0.05	1.26	0.56	**1.82**	m2
	smooth concrete	0.06	1.38	0.82	**2.20**	m2
	fibre cement	0.05	1.26	0.67	**1.93**	m2
	embossed or textured papered	0.05	1.26	0.56	**1.82**	m2

Unit Rates

		Man-Hours	Net Labour Price £	Net Mats Price £	Net Unit Price £	Unit
	cement rendered	0.07	1.60	0.82	**2.42**	m2
	fair face brickwork	0.07	1.74	0.93	**2.67**	m2
	fair face blockwork	0.08	1.86	1.08	**2.94**	m2
060	Walls in staircase areas over 300 mm wide:					
	plastered	0.05	1.24	0.56	**1.80**	m2
	smooth concrete	0.06	1.34	0.82	**2.16**	m2
	fibre cement	0.05	1.24	0.67	**1.91**	m2
	embossed or textured papered	0.05	1.24	0.56	**1.80**	m2
	cement rendered	0.07	1.55	0.82	**2.37**	m2
	fair face brickwork	0.07	1.69	0.93	**2.62**	m2
	fair face blockwork	0.08	1.84	1.08	**2.92**	m2
061	Ceilings over 300 mm wide:					
	plastered	0.05	1.26	0.56	**1.82**	m2
	smooth concrete	0.06	1.38	0.82	**2.20**	m2
	fibre cement	0.05	1.26	0.67	**1.93**	m2
	embossed or textured papered	0.05	1.26	0.56	**1.82**	m2
	cement rendered	0.07	1.60	0.82	**2.42**	m2
062	Ceilings 3.50 - 5.00 m high over 300 mm wide:					
	plastered	0.06	1.31	0.56	**1.87**	m2
	smooth concrete	0.06	1.43	0.82	**2.25**	m2
	fibre cement	0.06	1.31	0.67	**1.98**	m2
	embossed or textured papered	0.06	1.31	0.56	**1.87**	m2
	cement rendered	0.07	1.69	0.82	**2.51**	m2
063	Ceilings in staircase areas over 300 mm wide:					
	plastered	0.05	1.29	0.56	**1.85**	m2
	smooth concrete	0.06	1.41	0.82	**2.23**	m2
	fibre cement	0.05	1.29	0.67	**1.96**	m2
	embossed or textured papered	0.05	1.29	0.56	**1.85**	m2
	cement rendered	0.07	1.62	0.82	**2.44**	m2
	ACRYLATED RUBBER PAINT; UNDERCOAT - SPRAYED INTERNALLY					
064	**One coat of acrylated rubber, paint to primed surfaces; applied to dry film thickness of 55 microns**					
065	Walls over 300 mm wide:					
	plastered	0.06	1.38	5.24	**6.62**	m2
	smooth concrete	0.06	1.53	5.52	**7.05**	m2
	fibre cement	0.05	1.26	5.52	**6.78**	m2
	cement rendered	0.07	1.72	5.83	**7.55**	m2
	fair face brickwork	0.08	1.81	6.17	**7.98**	m2
	fair face blockwork	0.09	2.03	6.56	**8.59**	m2
066	Walls 3.50 - 5.00 m high where ceiling is of dissimilar finish; over 300 mm wide:					
	plastered	0.06	1.43	5.24	**6.67**	m2
	smooth concrete	0.07	1.60	5.52	**7.12**	m2

Unit Rates

		Man-Hours	Net Labour Price £	Net Mats Price £	Net Unit Price £	Unit
	fibre cement	0.06	1.31	5.52	**6.83**	m2
	cement rendered	0.08	1.81	5.83	**7.64**	m2
	fair face brickwork	0.08	1.88	6.17	**8.05**	m2
	fair face blockwork	0.09	2.15	6.56	**8.71**	m2
067	Walls in staircase areas over 300 mm wide:					
	plastered	0.06	1.41	5.24	**6.65**	m2
	smooth concrete	0.07	1.55	5.52	**7.07**	m2
	fibre cement	0.05	1.29	5.52	**6.81**	m2
	cement rendered	0.07	1.76	5.83	**7.59**	m2
	fair face brickwork	0.08	1.86	6.17	**8.03**	m2
	fair face blockwork	0.09	2.10	6.56	**8.66**	m2
068	Ceilings over 300 mm wide:					
	plastered	0.06	1.43	5.24	**6.67**	m2
	smooth concrete	0.07	1.60	5.52	**7.12**	m2
	fibre cement	0.06	1.31	5.52	**6.83**	m2
	cement rendered	0.08	1.81	5.82	**7.63**	m2
069	Ceilings 3.50 - 5.00 m high over 300 mm wide:					
	plastered	0.06	1.53	5.24	**6.77**	m2
	smooth concrete	0.07	1.69	5.52	**7.21**	m2
	fibre cement	0.06	1.38	5.52	**6.90**	m2
	cement rendered	0.08	1.88	5.82	**7.70**	m2
070	Ceilings in staircase areas over 300 mm wide:					
	plastered	0.06	1.45	5.24	**6.69**	m2
	smooth concrete	0.07	1.62	5.52	**7.14**	m2
	fibre cement	0.06	1.34	5.52	**6.86**	m2
	cement rendered	0.08	1.86	5.82	**7.68**	m2
	ACRYLATED RUBBER PAINT; FINISH - SPRAYED INTERNALLY					
071	**One coat of acrylated rubber paint gloss finish, to coated surfaces; applied to a dry film thickness of 50 microns**					
072	Walls over 300 mm wide:					
	plastered	0.04	1.03	3.37	**4.40**	m2
	smooth concrete	0.05	1.14	3.43	**4.57**	m2
	fibre cement	0.04	1.03	3.43	**4.46**	m2
	cement rendered	0.05	1.29	3.49	**4.78**	m2
	fair face brickwork	0.06	1.43	3.49	**4.92**	m2
	fair face blockwork	0.06	1.53	3.55	**5.08**	m2
073	Walls 3.50 - 5.00 m high where ceiling is of dissimilar finish; over 300 mm wide:					
	plastered	0.05	1.10	3.37	**4.47**	m2
	smooth concrete	0.05	1.19	3.43	**4.62**	m2
	fibre cement	0.05	1.10	3.43	**4.53**	m2
	cement rendered	0.06	1.34	3.49	**4.83**	m2
	fair face brickwork	0.06	1.48	3.49	**4.97**	m2

Unit Rates

		Man-Hours	Net Labour Price £	Net Mats Price £	Net Unit Price £	Unit
	fair face blockwork	0.07	1.55	3.55	**5.10**	m2
074	Walls in staircase areas over 300 mm wide:					
	plastered	0.04	1.05	3.37	**4.42**	m2
	smooth concrete	0.05	1.17	3.43	**4.60**	m2
	fibre cement	0.04	1.05	3.43	**4.48**	m2
	cement rendered	0.06	1.31	3.49	**4.80**	m2
	fair face brickwork	0.06	1.45	3.49	**4.94**	m2
	fair face blockwork	0.07	1.55	3.55	**5.10**	m2
075	Ceilings over 300 mm wide:					
	plastered	0.05	1.10	3.38	**4.48**	m2
	smooth concrete	0.05	1.19	3.44	**4.63**	m2
	fibre cement	0.05	1.10	3.44	**4.54**	m2
	cement rendered	0.06	1.34	3.49	**4.83**	m2
076	Ceilings 3.50 - 5.00 m high over 300 mm wide:					
	plastered	0.05	1.14	3.38	**4.52**	m2
	smooth concrete	0.05	1.26	3.44	**4.70**	m2
	fibre cement	0.05	1.14	3.44	**4.58**	m2
	cement rendered	0.06	1.41	3.49	**4.90**	m2
077	Ceilings in staircase areas over 300 mm wide:					
	plastered	0.05	1.12	3.38	**4.50**	m2
	smooth concrete	0.05	1.24	3.44	**4.68**	m2
	fibre cement	0.05	1.12	3.44	**4.56**	m2
	cement rendered	0.06	1.38	3.49	**4.87**	m2
	Note 078 - 079 not used					
	MULTICOLOUR FINISH - SPRAYED					
080	**One primer/sealer bonding coat (to receive multicolour finish)**					
081	Walls over 300 mm wide:					
	plastered	0.07	1.67	0.79	**2.46**	m2
	smooth concrete	0.07	1.67	0.79	**2.46**	m2
	fibre cement	0.07	1.55	0.79	**2.34**	m2
	cement rendered	0.08	1.91	0.83	**2.74**	m2
	fair face brickwork	0.09	2.15	0.88	**3.03**	m2
	fair face blockwork	0.10	2.38	0.88	**3.26**	m2
082	Walls 3.50 - 5.00 m high where ceiling is of dissimilar finish over 300 mm wide:					
	plastered	0.08	1.91	0.79	**2.70**	m2
	smooth concrete	0.08	1.91	0.79	**2.70**	m2
	fibre cement	0.08	1.79	0.79	**2.58**	m2
	cement rendered	0.09	2.15	0.83	**2.98**	m2
	fair face brickwork	0.10	2.38	0.88	**3.26**	m2
	fair face blockwork	0.11	2.62	0.88	**3.50**	m2

Unit Rates

		Man-Hours	Net Labour Price £	Net Mats Price £	Net Unit Price £	Unit
083	Walls in staircase areas over 300 mm wide:					
	plastered	0.08	1.91	0.79	**2.70**	m2
	smooth concrete	0.08	1.91	0.79	**2.70**	m2
	fibre cement	0.08	1.79	0.79	**2.58**	m2
	cement rendered	0.09	2.15	0.83	**2.98**	m2
	fair face brickwork	0.10	2.38	0.88	**3.26**	m2
	fair face blockwork	0.11	2.62	0.88	**3.50**	m2
084	**One additional primer/sealer bonding coat (to receive multicolour finish)**					
085	Walls over 300 mm wide:					
	cement rendered	0.08	1.91	0.79	**2.70**	m2
	fair face brickwork	0.09	2.15	0.79	**2.94**	m2
	fair face blockwork	0.10	2.38	0.79	**3.17**	m2
086	Walls 3.50 - 5.00 m high where ceiling is of dissimilar finish over 300 mm wide:					
	cement rendered	0.09	2.15	0.79	**2.94**	m2
	fair face brickwork	0.10	2.38	0.79	**3.17**	m2
	fair face blockwork	0.11	2.62	0.79	**3.41**	m2
087	Walls in staircase areas over 300 mm wide:					
	cement rendered	0.09	2.15	0.79	**2.94**	m2
	fair face brickwork	0.10	2.38	0.79	**3.17**	m2
	fair face blockwork	0.11	2.62	0.79	**3.41**	m2
088	**One coat multicolour finish to primed surfaces**					
089	Walls over 300 mm wide:					
	plastered	0.12	2.86	2.45	**5.31**	m2
	smooth concrete	0.12	2.86	2.45	**5.31**	m2
	fibre cement	0.12	2.86	2.45	**5.31**	m2
	cement rendered	0.14	3.34	3.44	**6.78**	m2
	fair face brickwork	0.15	3.58	3.93	**7.51**	m2
	fair face blockwork	0.16	3.82	4.42	**8.24**	m2
090	Walls 3.50 - 5.00 m high where ceiling is of dissimilar finish over 300 mm wide:					
	plastered	0.13	3.10	2.45	**5.55**	m2
	smooth concrete	0.13	3.10	2.45	**5.55**	m2
	fibre cement	0.13	3.10	2.45	**5.55**	m2
	cement rendered	0.15	3.58	3.44	**7.02**	m2
	fair face brickwork	0.16	3.82	3.93	**7.75**	m2
	fair face blockwork	0.17	4.05	4.42	**8.47**	m2
091	Walls in staircase areas over 300 mm wide:					
	plastered	0.13	3.10	2.45	**5.55**	m2
	smooth concrete	0.13	3.10	2.45	**5.55**	m2
	fibre cement	0.13	3.10	2.45	**5.55**	m2
	cement rendered	0.15	3.58	3.44	**7.02**	m2
	fair face brickwork	0.16	3.82	3.93	**7.75**	m2
	fair face blockwork	0.17	4.05	4.42	**8.47**	m2

Unit Rates

		Man-Hours	Net Labour Price £	Net Mats Price £	Net Unit Price £	Unit
	Note 092 - 099 not used					
	NEW WORK INTERNALLY - FLOOR PAINTS					
100	**One coat semi-gloss floor paint**					
101	First coat to unprimed surfaces:					
	over 300 mm girth	0.10	1.45	0.39	**1.84**	m2
	not exceeding 150 mm girth	0.02	0.30	0.06	**0.36**	m
	150 - 300 mm girth	0.04	0.53	0.12	**0.65**	m
102	Second and subsequent coats:					
	over 300 mm girth	0.09	1.37	0.32	**1.69**	m2
	not exceeding 150 mm girth	0.02	0.30	0.05	**0.35**	m
	150 - 300 mm girth	0.04	0.53	0.10	**0.63**	m
103	**Prepare; one coat quick drying Thermoplastic Protective floor paint thinned by 10% with clean water**					
104	Floor, concrete or the like:					
	over 300 mm girth	0.10	1.52	0.40	**1.92**	m2
	not exceeding 150 mm girth	0.02	0.30	0.06	**0.36**	m
	150 - 300 mm girth	0.04	0.53	0.12	**0.65**	m
105	**One coat quick drying floor paint and each subsequent coat**					
106	Floor, concrete or the like:					
	over 300 mm girth	0.10	1.52	0.52	**2.04**	m2
	not exceeding 150 mm girth	0.02	0.30	0.08	**0.38**	m
	150 - 300 mm girth	0.04	0.53	0.16	**0.69**	m
107	**One coat anti-slip floor paint with bauxite aggregate and each subsequent coat on floor paint**					
108	Floor, concrete or the like:					
	over 300 mm girth	0.10	1.52	0.72	**2.24**	m2
	not exceeding 150 mm girth	0.02	0.30	1.04	**1.34**	m
	150 - 300 mm girth	0.04	0.53	0.22	**0.75**	m
VD	**METALWORK - INTERNALLY**					
	METALWORK PRIMERS - INTERNALLY					
001	**Prepare; one coat two-pack etching primer, metalwork surfaces, applied to a dry film thickness of 3 microns**					
002	General surfaces:					
	over 300 mm girth	0.22	3.35	0.59	**3.94**	m2
	not exceeding 150 mm girth	0.07	1.01	0.10	**1.11**	m
	150 - 300 mm girth	0.10	1.51	0.21	**1.72**	m

Unit Rates

		Man-Hours	Net Labour Price £	Net Mats Price £	Net Unit Price £	Unit
	isolated; not exceeding 0.50 m2	0.16	2.44	0.31	**2.75**	Nr
003	Glazed doors and screens in panes:					
	small - not exceeding 0.10 m2	0.55	8.38	0.33	**8.71**	m2
	medium - 0.10 - 0.50 m2	0.39	5.86	0.28	**6.14**	m2
	large - 0.50 - 1.00 m2	0.36	5.53	0.23	**5.76**	m2
	extra large - over 1.00 m2	0.33	5.03	0.23	**5.26**	m2
004	Windows in panes:					
	small - not exceeding 0.10 m2	0.55	8.38	0.28	**8.66**	m2
	medium - 0.10 - 0.50 m2	0.39	5.86	0.23	**6.09**	m2
	large - 0.50 - 1.00 m2	0.36	5.53	0.23	**5.76**	m2
	extra large - over 1.00 m2	0.33	5.03	0.18	**5.21**	m2
005	Structural members:					
	over 300 mm girth	0.25	3.85	0.59	**4.44**	m2
	not exceeding 150 mm girth	0.08	1.16	0.10	**1.26**	m
	150 - 300 mm girth	0.11	1.74	0.21	**1.95**	m
006	Members of roof trusses:					
	over 300 mm girth	0.33	5.03	0.74	**5.77**	m2
	not exceeding 150 mm girth	0.10	1.51	0.10	**1.61**	m
	150 - 300 mm girth	0.15	2.27	0.26	**2.53**	m
007	Radiators:					
	over 300 mm girth	0.29	4.36	0.64	**5.00**	m2
008	Pipes and conduits, ducting, trunking and the like:					
	over 300 mm girth	0.24	3.69	0.69	**4.38**	m2
	not exceeding 150 mm girth	0.07	1.11	0.10	**1.21**	m
	150 - 300 mm girth	0.11	1.66	0.21	**1.87**	m
009	Staircases:					
	over 300 mm girth	0.24	3.69	0.59	**4.28**	m2
	not exceeding 150 mm girth	0.07	1.11	0.10	**1.21**	m
	150 - 300 mm girth	0.11	1.66	0.21	**1.87**	m
010	**Prepare; one coat zinc phosphate primer, metalwork surfaces**					
011	General surfaces:					
	over 300 mm girth	0.22	3.35	0.95	**4.30**	m2
	not exceeding 150 mm girth	0.07	1.01	0.14	**1.15**	m
	150 - 300 mm girth	0.10	1.51	0.28	**1.79**	m
	isolated; not exceeding 0.50 m2	0.16	2.44	0.46	**2.90**	Nr
012	Glazed doors and screens in panes:					
	small - not exceeding 0.10 m2	0.55	8.38	0.49	**8.87**	m2
	medium - 0.10 - 0.50 m2	0.39	5.86	0.39	**6.25**	m2
	large - 0.50 - 1.00 m2	0.36	5.53	0.34	**5.87**	m2
	extra large - over 1.00 m2	0.33	5.03	0.30	**5.33**	m2

Unit Rates

INTERNAL WORK

		Man-Hours	Net Labour Price £	Net Mats Price £	Net Unit Price £	Unit
013	Windows in panes:					
	small - not exceeding 0.10 m2	0.55	8.38	0.39	**8.77**	m2
	medium - 0.10 - 0.50 m2	0.39	5.86	0.34	**6.20**	m2
	large - 0.50 - 1.00 m2	0.36	5.53	0.30	**5.83**	m2
	extra large - over 1.00 m2	0.33	5.03	0.25	**5.28**	m2
014	Structural members:					
	over 300 mm girth	0.25	3.85	0.95	**4.80**	m2
	not exceeding 150 mm girth	0.08	1.16	0.14	**1.30**	m
	150 - 300 mm girth	0.11	1.74	0.28	**2.02**	m
015	Members of roof trusses:					
	over 300 mm girth	0.33	5.03	1.23	**6.26**	m2
	not exceeding 150 mm girth	0.10	1.51	0.18	**1.69**	m
	150 - 300 mm girth	0.15	2.27	0.36	**2.63**	m
016	Radiators:					
	over 300 mm girth	0.29	4.36	1.07	**5.43**	m2
017	Pipes and conduits, ducting, trunking and the like:					
	over 300 mm girth	0.24	3.69	1.00	**4.69**	m2
	not exceeding 150 mm girth	0.07	1.11	0.15	**1.26**	m
	150 - 300 mm girth	0.11	1.66	0.29	**1.95**	m
018	Staircases:					
	over 300 mm girth	0.24	3.69	0.95	**4.64**	m2
	not exceeding 150 mm girth	0.07	1.11	0.14	**1.25**	m
	150 - 300 mm girth	0.11	1.66	0.28	**1.94**	m
019	**Prepare; one coat zinc chromate primer, metalwork surfaces**					
020	General surfaces:					
	over 300 mm girth	0.22	3.35	0.92	**4.27**	m2
	not exceeding 150 mm girth	0.07	1.01	0.13	**1.14**	m
	150 - 300 mm girth	0.10	1.51	0.27	**1.78**	m
	isolated; not exceeding 0.50 m2	0.16	2.44	0.45	**2.89**	Nr
021	Glazed doors and screens in panes:					
	small - not exceeding 0.10 m2	0.55	8.38	0.47	**8.85**	m2
	medium - 0.10 - 0.50 m2	0.39	5.86	0.38	**6.24**	m2
	large - 0.50 - 1.00 m2	0.36	5.53	0.33	**5.86**	m2
	extra large - over 1.00 m2	0.33	5.03	0.29	**5.32**	m2
022	Windows in panes:					
	small - not exceeding 0.10 m2	0.55	8.38	0.38	**8.76**	m2
	medium - 0.10 - 0.50 m2	0.39	5.86	0.33	**6.19**	m2
	large - 0.50 - 1.00 m2	0.36	5.53	0.29	**5.82**	m2
	extra large - over 1.00 m2	0.33	5.03	0.25	**5.28**	m2

Unit Rates

INTERNAL WORK VD

		Man-Hours	Net Labour Price £	Net Mats Price £	Net Unit Price £	Unit
023	Structural members:					
	over 300 mm girth	0.25	3.85	0.92	**4.77**	m2
	not exceeding 150 mm girth	0.08	1.16	0.13	**1.29**	m
	150 - 300 mm girth	0.11	1.74	0.27	**2.01**	m
024	Members of roof trusses:					
	over 300 mm girth	0.33	5.03	1.20	**6.23**	m2
	not exceeding 150 mm girth	0.10	1.51	0.18	**1.69**	m
	150 - 300 mm girth	0.15	2.27	0.35	**2.62**	m
025	Radiators:					
	over 300 mm girth	0.29	4.36	1.04	**5.40**	m2
026	Pipes and conduits, ducting, trunking and the like:					
	over 300 mm girth	0.24	3.69	0.97	**4.66**	m2
	not exceeding 150 mm girth	0.07	1.11	0.14	**1.25**	m
	150 - 300 mm girth	0.11	1.66	0.29	**1.95**	m
027	Staircases:					
	over 300 mm girth	0.24	3.69	0.92	**4.61**	m2
	not exceeding 150 mm girth	0.07	1.11	0.13	**1.24**	m
	150 - 300 mm girth	0.11	1.66	0.27	**1.93**	m
028	**Prepare; one coat of zinc rich primer, metalwork surfaces**					
029	General surfaces:					
	over 300 mm girth	0.22	3.35	4.54	**7.89**	m2
	not exceeding 150 mm girth	0.07	1.01	0.67	**1.68**	m
	150 - 300 mm girth	0.10	1.51	1.36	**2.87**	m
	isolated; not exceeding 0.50 m2	0.16	2.44	2.27	**4.71**	Nr
030	Glazed doors and screens in panes:					
	small - not exceeding 0.10 m2	0.55	8.38	2.29	**10.67**	m2
	medium - 0.10 - 0.50 m2	0.39	5.86	1.83	**7.69**	m2
	large - 0.50 - 1.00 m2	0.36	5.53	1.60	**7.13**	m2
	extra large - over 1.00 m2	0.33	5.03	1.39	**6.42**	m2
031	Windows in panes:					
	small - not exceeding 0.10 m2	0.55	8.38	1.83	**10.21**	m2
	medium - 0.10 - 0.50 m2	0.39	5.86	1.60	**7.46**	m2
	large - 0.50 - 1.00 m2	0.36	5.53	1.39	**6.92**	m2
	extra large - over 1.00 m2	0.33	5.03	1.16	**6.19**	m2
032	Structural members:					
	over 300 mm girth	0.25	3.85	4.54	**8.39**	m2
	not exceeding 150 mm girth	0.08	1.16	0.67	**1.83**	m
	150 - 300 mm girth	0.11	1.74	1.36	**3.10**	m
033	Members of roof trusses:					
	over 300 mm girth	0.33	5.03	5.94	**10.97**	m2

Unit Rates

		Man-Hours	Net Labour Price £	Net Mats Price £	Net Unit Price £	Unit
	not exceeding 150 mm girth	0.10	1.52	0.88	**2.40**	m
	150 - 300 mm girth	0.15	2.27	1.77	**4.04**	m
034	Radiators:					
	over 300 mm girth	0.29	4.36	5.15	**9.51**	m2
035	Pipes and conduits, ducting, trunking and the like:					
	over 300 mm girth	0.24	3.69	4.83	**8.52**	m2
	not exceeding 150 mm girth	0.07	1.11	0.73	**1.84**	m
	150 - 300 mm girth	0.11	1.66	1.44	**3.10**	m
036	Staircases:					
	over 300 mm girth	0.24	3.69	4.54	**8.23**	m2
	not exceeding 150 mm girth	0.07	1.11	0.67	**1.78**	m
	150 - 300 mm girth	0.11	1.66	1.36	**3.02**	m
037	**Prepare; one coat red oxide primer, metalwork surfaces**					
038	General surfaces:					
	over 300 mm girth	0.22	3.35	0.94	**4.29**	m2
	not exceeding 150 mm girth	0.07	1.01	0.14	**1.15**	m
	150 - 300 mm girth	0.10	1.51	0.28	**1.79**	m
	isolated; not exceeding 0.50 m2	0.16	2.44	0.46	**2.90**	Nr
039	Glazed doors and screens in panes:					
	small - not exceeding 0.10 m2	0.55	8.38	0.48	**8.86**	m2
	medium - 0.10 - 0.50 m2	0.39	5.86	0.39	**6.25**	m2
	large - 0.50 - 1.00 m2	0.36	5.53	0.34	**5.87**	m2
	extra large - over 1.00 m2	0.33	5.03	0.30	**5.33**	m2
040	Windows in panes:					
	small - not exceeding 0.10 m2	0.55	8.38	0.39	**8.77**	m2
	medium - 0.10 - 0.50 m2	0.39	5.86	0.34	**6.20**	m2
	large - 0.50 - 1.00 m2	0.36	5.53	0.30	**5.83**	m2
	extra large - over 1.00 m2	0.33	5.03	0.25	**5.28**	m2
041	Structural members:					
	over 300 mm girth	0.25	3.85	0.94	**4.79**	m2
	not exceeding 150 mm girth	0.08	1.16	0.14	**1.30**	m
	150 - 300 mm girth	0.11	1.74	0.28	**2.02**	m
042	Members of roof trusses:					
	over 300 mm girth	0.33	5.03	1.23	**6.26**	m2
	not exceeding 150 mm girth	0.10	1.51	0.18	**1.69**	m
	150 - 300 mm girth	0.15	2.27	0.36	**2.63**	m
043	Radiators:					
	over 300 mm girth	0.29	4.36	1.07	**5.43**	m2

Unit Rates

		Man-Hours	Net Labour Price £	Net Mats Price £	Net Unit Price £	Unit
044	Pipes and conduits, ducting, trunking and the like:					
	over 300 mm girth	0.24	3.69	1.00	**4.69**	m2
	not exceeding 150 mm girth	0.07	1.11	0.15	**1.26**	m
	150 - 300 mm girth	0.11	1.66	0.29	**1.95**	m
045	Staircases:					
	over 300 mm girth	0.24	3.69	0.94	**4.63**	m2
	not exceeding 150 mm girth	0.07	1.11	0.14	**1.25**	m
	150 - 300 mm girth	0.11	1.66	0.28	**1.94**	m
046	**Prepare; one coat of galvanising paint, metalwork surfaces, applied to a dry film thickness of 125 microns**					
047	General surfaces:					
	over 300 mm girth	0.19	2.89	1.90	**4.79**	m2
	not exceeding 150 mm girth	0.06	0.87	0.31	**1.18**	m
	150 - 300 mm girth	0.09	1.31	0.62	**1.93**	m
	isolated; not exceeding 0.50 m2	0.12	1.83	1.04	**2.87**	Nr
048	Glazed doors and screens in panes:					
	small - not exceeding 0.10 m2	0.54	8.22	1.06	**9.28**	m2
	medium - 0.10 - 0.50 m2	0.37	5.64	0.85	**6.49**	m2
	large - 0.50 - 1.00 m2	0.35	5.39	0.74	**6.13**	m2
	extra large - over 1.00 m2	0.32	4.83	0.64	**5.47**	m2
049	Windows in panes:					
	small - not exceeding 0.10 m2	0.54	8.22	0.85	**9.07**	m2
	medium - 0.10 - 0.50 m2	0.37	5.64	0.74	**6.38**	m2
	large - 0.50 - 1.00 m2	0.35	5.39	0.64	**6.03**	m2
	extra large - over 1.00 m2	0.32	4.83	0.54	**5.37**	m2
050	Structural members:					
	over 300 mm girth	0.25	3.85	1.90	**5.75**	m2
	not exceeding 150 mm girth	0.08	1.16	0.31	**1.47**	m
	150 - 300 mm girth	0.11	1.74	0.62	**2.36**	m
051	Members of roof trusses:					
	over 300 mm girth	0.32	4.87	2.37	**7.24**	m2
	not exceeding 150 mm girth	0.10	1.55	0.39	**1.94**	m
	150 - 300 mm girth	0.15	2.35	0.77	**3.12**	m
052	Radiators:					
	over 300 mm girth	0.28	4.26	2.11	**6.37**	m2
053	Pipes and conduits, ducting, trunking and the like:					
	over 300 mm girth	0.21	3.18	2.23	**5.41**	m2
	not exceeding 150 mm girth	0.06	0.97	0.37	**1.34**	m
	150 - 300 mm girth	0.09	1.43	0.73	**2.16**	m

Unit Rates

		Man-Hours	Net Labour Price £	Net Mats Price £	Net Unit Price £	Unit
054	Staircases:					
	over 300 mm girth	0.25	3.85	1.90	**5.75**	m2
	not exceeding 150 mm girth	0.08	1.16	0.31	**1.47**	m
	150 - 300 mm girth	0.11	1.74	0.60	**2.34**	m
055	**One coat of galvanising paint; second and subsequent coats to previously coated metalwork surfaces; applied to a dry film thickness of 125 microns**					
056	General surfaces:					
	over 300 mm girth	0.18	2.74	1.58	**4.32**	m2
	not exceeding 150 mm girth	0.05	0.78	0.26	**1.04**	m
	150 - 300 mm girth	0.08	1.22	0.52	**1.74**	m
	isolated; not exceeding 0.50 m2	0.14	2.13	0.78	**2.91**	Nr
057	Glazed doors and screens in panes:					
	small - not exceeding 0.10 m2	0.53	8.04	0.80	**8.84**	m2
	medium - 0.10 - 0.50 m2	0.35	5.30	0.64	**5.94**	m2
	large - 0.50 - 1.00 m2	0.32	4.90	0.57	**5.47**	m2
	extra large - over 1.00 m2	0.28	4.29	0.64	**4.93**	m2
058	Windows in panes:					
	small - not exceeding 0.10 m2	0.53	8.04	0.64	**8.68**	m2
	medium - 0.10 - 0.50 m2	0.36	5.45	0.57	**6.02**	m2
	large - 0.50 - 1.00 m2	0.33	5.06	0.49	**5.55**	m2
	extra large - over 1.00 m2	0.29	4.45	0.41	**4.86**	m2
059	Structural members:					
	over 300 mm girth	0.23	3.44	1.58	**5.02**	m2
	not exceeding 150 mm girth	0.07	1.05	0.26	**1.31**	m
	150 - 300 mm girth	0.10	1.52	0.52	**2.04**	m
060	Members of roof trusses:					
	over 300 mm girth	0.32	4.87	1.97	**6.84**	m2
	not exceeding 150 mm girth	0.09	1.37	0.32	**1.69**	m
	150 - 300 mm girth	0.14	2.16	0.65	**2.81**	m
061	Radiators:					
	over 300 mm girth	0.28	4.25	1.76	**6.01**	m2
062	Pipes and conduits, ducting, trunking and the like:					
	over 300 mm girth	0.20	3.05	1.86	**4.91**	m2
	not exceeding 150 mm girth	0.05	0.72	0.30	**1.02**	m
	150 - 300 mm girth	0.08	1.19	0.61	**1.80**	m
063	Staircases:					
	over 300 mm girth	0.24	3.59	1.58	**5.17**	m2
	not exceeding 150 mm girth	0.07	1.05	0.26	**1.31**	m
	150 - 300 mm girth	0.10	1.52	0.52	**2.04**	m

Unit Rates

		Man-Hours	Net Labour Price £	Net Mats Price £	Net Unit Price £	Unit
	METALWORK UNDERCOATS - INTERNALLY					
064	**One undercoat alkyd based paint; primed metalwork surfaces**					
065	General surfaces:					
	over 300 mm girth	0.17	2.51	0.52	**3.03**	m2
	not exceeding 150 mm girth	0.05	0.76	0.07	**0.83**	m
	150 - 300 mm girth	0.08	1.14	0.15	**1.29**	m
	isolated; not exceeding 0.50 m2	0.16	2.44	0.24	**2.68**	Nr
066	Glazed doors and screens in panes:					
	small - not exceeding 0.10 m2	0.50	7.54	0.27	**7.81**	m2
	medium - 0.10 - 0.50 m2	0.33	5.03	0.22	**5.25**	m2
	large - 0.50 - 1.00 m2	0.31	4.69	0.20	**4.89**	m2
	extra large - over 1.00 m2	0.28	4.19	0.17	**4.36**	m2
067	Windows in panes:					
	small - not exceeding 0.10 m2	0.50	7.54	0.22	**7.76**	m2
	medium - 0.10 - 0.50 m2	0.33	5.03	0.20	**5.23**	m2
	large - 0.50 - 1.00 m2	0.31	4.69	0.17	**4.86**	m2
	extra large - over 1.00 m2	0.28	4.19	0.15	**4.34**	m2
068	Structural members:					
	over 300 mm girth	0.22	3.35	0.52	**3.87**	m2
	not exceeding 150 mm girth	0.07	1.01	0.07	**1.08**	m
	150 - 300 mm girth	0.10	1.51	0.15	**1.66**	m
069	Members of roof trusses:					
	over 300 mm girth	0.30	4.52	0.68	**5.20**	m2
	not exceeding 150 mm girth	0.09	1.36	0.10	**1.46**	m
	150 - 300 mm girth	0.13	2.04	0.20	**2.24**	m
070	Radiators:					
	over 300 mm girth	0.26	4.02	0.57	**4.59**	m2
071	Pipes and conduits, ducting, trunking and the like:					
	over 300 mm girth	0.18	2.77	0.57	**3.34**	m2
	not exceeding 150 mm girth	0.06	0.84	0.08	**0.92**	m
	150 - 300 mm girth	0.08	1.23	0.16	**1.39**	m
072	Staircases:					
	over 300 mm girth	0.22	3.35	0.52	**3.87**	m2
	not exceeding 150 mm girth	0.07	1.01	0.07	**1.08**	m
	150 - 300 mm girth	0.10	1.51	0.15	**1.66**	m

Unit Rates

METALWORK GLOSS FINISH - INTERNALLY

		Man-Hours	Net Labour Price £	Net Mats Price £	Net Unit Price £	Unit
073	**One coat alkyd based paint, gloss finish; undercoated metalwork surfaces**					
074	General surfaces:					
	over 300 mm girth	0.19	2.85	0.52	**3.37**	m2
	not exceeding 150 mm girth	0.06	0.85	0.07	**0.92**	m
	150 - 300 mm girth	0.09	1.29	0.15	**1.44**	m
	isolated; not exceeding 0.50 m2	0.16	2.44	0.25	**2.69**	Nr
075	Glazed doors and screens in panes:					
	small - not exceeding 0.10 m2	0.52	7.87	0.27	**8.14**	m2
	medium - 0.10 - 0.50 m2	0.35	5.36	0.22	**5.58**	m2
	large - 0.50 - 1.00 m2	0.33	5.03	0.20	**5.23**	m2
	extra large - over 1.00 m2	0.30	4.52	0.17	**4.69**	m2
076	Windows in panes:					
	small - not exceeding 0.10 m2	0.52	7.87	0.22	**8.09**	m2
	medium - 0.10 - 0.50 m2	0.35	5.36	0.20	**5.56**	m2
	large - 0.50 - 1.00 m2	0.33	5.03	0.17	**5.20**	m2
	extra large - over 1.00 m2	0.30	4.52	0.15	**4.67**	m2
077	Structural members:					
	over 300 mm girth	0.24	3.69	0.52	**4.21**	m2
	not exceeding 150 mm girth	0.07	1.11	0.07	**1.18**	m
	150 - 300 mm girth	0.11	1.66	0.15	**1.81**	m
078	Members of roof trusses:					
	over 300 mm girth	0.31	4.69	0.68	**5.37**	m2
	not exceeding 150 mm girth	0.09	1.40	0.09	**1.49**	m
	150 - 300 mm girth	0.14	2.12	0.20	**2.32**	m
079	Radiators:					
	over 300 mm girth	0.28	4.19	0.57	**4.76**	m2
080	Pipes and conduits, ducting, trunking and the like:					
	over 300 mm girth	0.21	3.14	0.57	**3.71**	m2
	not exceeding 150 mm girth	0.06	0.94	0.08	**1.02**	m
	150 - 300 mm girth	0.09	1.40	0.15	**1.55**	m
081	Staircases:					
	over 300 mm girth	0.23	3.52	0.52	**4.04**	m2
	not exceeding 150 mm girth	0.07	1.05	0.07	**1.12**	m
	150 - 300 mm girth	0.11	1.60	0.15	**1.75**	m

	Unit Rates	Man-Hours	Net Labour Price £	Net Mats Price £	Net Unit Price £	Unit
VE	**METALWORK - INTERNALLY**					
	METALWORK ACRYLATED RUBBER - INTERNALLY					
001	**Prepare; one coat acrylated rubber paint primer, metalwork surfaces, applied to a dry film thickness of 55 microns**					
002	General surfaces:					
	over 300 mm girth	0.25	3.85	1.35	**5.20**	m2
	not exceeding 150 mm girth	0.08	1.16	0.22	**1.38**	m
	150 - 300 mm girth	0.11	1.74	0.44	**2.18**	m
	isolated; not exceeding 0.50 m2	0.16	2.44	0.73	**3.17**	Nr
003	Glazed doors and screens in panes:					
	small - not exceeding 0.10 m2	0.63	9.64	0.75	**10.39**	m2
	medium - 0.10 - 0.50 m2	0.44	6.75	0.61	**7.36**	m2
	large - 0.50 - 1.00 m2	0.42	6.37	0.53	**6.90**	m2
	extra large - over 1.00 m2	0.38	5.79	0.46	**6.25**	m2
004	Windows in panes:					
	small - not exceeding 0.10 m2	0.63	9.64	0.61	**10.25**	m2
	medium - 0.10 - 0.50 m2	0.44	6.75	0.53	**7.28**	m2
	large - 0.50 - 1.00 m2	0.42	6.37	0.46	**6.83**	m2
	extra large - over 1.00 m2	0.38	5.79	0.39	**6.18**	m2
005	Structural members:					
	over 300 mm girth	0.29	4.45	1.35	**5.80**	m2
	not exceeding 150 mm girth	0.09	1.33	0.22	**1.55**	m
	150 - 300 mm girth	0.13	1.98	0.44	**2.42**	m
006	Members of roof trusses:					
	over 300 mm girth	0.38	5.79	1.68	**7.47**	m2
	not exceeding 150 mm girth	0.11	1.74	0.27	**2.01**	m
	150 - 300 mm girth	0.17	2.60	0.55	**3.15**	m
007	Radiators:					
	over 300 mm girth	0.33	5.01	1.50	**6.51**	m2
008	Pipes and conduits, ducting, trunking and the like:					
	over 300 mm girth	0.28	4.23	1.58	**5.81**	m2
	not exceeding 150 mm girth	0.08	1.28	0.26	**1.54**	m
	150 - 300 mm girth	0.16	2.41	0.52	**2.93**	m
009	Staircases:					
	over 300 mm girth	0.28	4.23	1.35	**5.58**	m2
	not exceeding 150 mm girth	0.08	1.28	0.22	**1.50**	m
	150 - 300 mm girth	0.13	1.90	0.44	**2.34**	m

INTERNAL WORK VE

	Unit Rates	Man-Hours	Net Labour Price £	Net Mats Price £	Net Unit Price £	Unit
010	One undercoat acrylated rubber paint; primed metal-work surfaces; applied to a dry film thickness of 100 microns					
011	General surfaces:					
	over 300 mm girth	0.19	2.89	9.07	**11.96**	m2
	not exceeding 150 mm girth	0.06	0.87	1.49	**2.36**	m
	150 - 300 mm girth	0.09	1.31	2.99	**4.30**	m
	isolated; not exceeding 0.50 m2	0.16	2.44	1.10	**3.54**	Nr
012	Glazed doors and screens in panes:					
	small - not exceeding 0.10 m2	0.57	8.68	5.00	**13.68**	m2
	medium - 0.10 - 0.50 m2	0.38	5.79	4.01	**9.80**	m2
	large - 0.50 - 1.00 m2	0.35	5.39	3.51	**8.90**	m2
	extra large - over 1.00 m2	0.32	4.83	3.01	**7.84**	m2
013	Windows in panes:					
	small - not exceeding 0.10 m2	0.57	8.68	4.01	**12.69**	m2
	medium - 0.10 - 0.50 m2	0.38	5.79	3.51	**9.30**	m2
	large - 0.50 - 1.00 m2	0.35	5.39	3.01	**8.40**	m2
	extra large - over 1.00 m2	0.32	4.83	2.51	**7.34**	m2
014	Structural members:					
	over 300 mm girth	0.25	3.85	9.07	**12.92**	m2
	not exceeding 150 mm girth	0.08	1.16	1.49	**2.65**	m
	150 - 300 mm girth	0.11	1.74	2.99	**4.73**	m
015	Members of roof trusses:					
	over 300 mm girth	0.34	5.21	10.51	**15.72**	m2
	not exceeding 150 mm girth	0.10	1.55	1.73	**3.28**	m
	150 - 300 mm girth	0.15	2.35	3.46	**5.81**	m
016	Radiators:					
	over 300 mm girth	0.30	4.63	9.98	**14.61**	m2
017	Pipes and conduits, ducting, trunking and the like:					
	over 300 mm girth	0.21	3.18	9.98	**13.16**	m2
	not exceeding 150 mm girth	0.06	0.97	1.64	**2.61**	m
	150 - 300 mm girth	0.09	1.43	3.29	**4.72**	m
018	Staircases:					
	over 300 mm girth	0.25	3.85	9.07	**12.92**	m2
	not exceeding 150 mm girth	0.08	1.16	1.49	**2.65**	m
	150 - 300 mm girth	0.11	1.74	2.99	**4.73**	m
019	One coat acrylated rubber paint, gloss finish; undercoated metalwork surfaces; applied to a dry film thickness of 25 microns					
020	General surfaces:					
	over 300 mm girth	0.22	3.29	2.40	**5.69**	m2
	not exceeding 150 mm girth	0.07	0.99	0.36	**1.35**	m

Unit Rates

		Man-Hours	Net Labour Price £	Net Mats Price £	Net Unit Price £	Unit
	150 - 300 mm girth	0.10	1.49	0.79	**2.28**	m
	isolated; not exceeding 0.50 m2	0.16	2.44	1.31	**3.75**	Nr
021	Glazed doors and screens in panes:					
	small - not exceeding 0.10 m2	0.60	9.06	1.33	**10.39**	m2
	medium - 0.10 - 0.50 m2	0.41	6.17	1.07	**7.24**	m2
	large - 0.50 - 1.00 m2	0.38	5.79	0.94	**6.73**	m2
	extra large - over 1.00 m2	0.34	5.21	0.81	**6.02**	m2
022	Windows in panes:					
	small - not exceeding 0.10 m2	0.60	9.06	1.07	**10.13**	m2
	medium - 0.10 - 0.50 m2	0.41	6.17	0.94	**7.11**	m2
	large - 0.50 - 1.00 m2	0.38	5.79	0.85	**6.64**	m2
	extra large - over 1.00 m2	0.34	5.21	0.68	**5.89**	m2
023	Structural members:					
	over 300 mm girth	0.28	4.23	2.40	**6.63**	m2
	not exceeding 150 mm girth	0.08	1.28	0.39	**1.67**	m
	150 - 300 mm girth	0.13	1.90	0.79	**2.69**	m
024	Members of roof trusses:					
	over 300 mm girth	0.35	5.39	3.00	**8.39**	m2
	not exceeding 150 mm girth	0.11	1.63	0.49	**2.12**	m
	150 - 300 mm girth	0.16	2.44	0.99	**3.43**	m
025	Radiators:					
	over 300 mm girth	0.32	4.83	2.66	**7.49**	m2
026	Pipes, conduits, ducting, trunking and the like:					
	over 300 mm girth	0.24	3.61	2.82	**6.43**	m2
	not exceeding 150 mm girth	0.07	1.07	0.46	**1.53**	m
	150 - 300 mm girth	0.11	1.63	0.93	**2.56**	m
027	Staircases:					
	over 300 mm girth	0.27	4.05	2.40	**6.45**	m2
	not exceeding 150 mm girth	0.08	1.20	0.39	**1.59**	m
	150 - 300 mm girth	0.12	1.83	0.79	**2.62**	m
	METALWORK ALUMINIUM PAINT - INTERNALLY					
028	**Prepare; one coat aluminium paint, metalwork surfaces**					
029	Boilers, tanks, calorifiers and the like:					
	over 300 mm girth	0.26	4.02	1.16	**5.18**	m2
030	Pipes, conduits, ducting, trunking and the like:					
	over 300 mm girth	0.18	2.77	1.27	**4.04**	m2
	not exceeding 150 mm girth	0.06	0.84	0.18	**1.02**	m
	150 - 300 mm girth	0.08	1.23	0.38	**1.61**	m

Unit Rates

		Man-Hours	Net Labour Price £	Net Mats Price £	Net Unit Price £	Unit
	METALWORK SPECIAL FINISHES					
031	**Prepare; one coat hammered metal finish, unprimed metalwork surfaces**					
032	General surfaces:					
	over 300 mm girth	0.19	2.89	3.34	**6.23**	m2
	not exceeding 150 mm girth	0.06	0.87	0.55	**1.42**	m
	150 - 300 mm girth	0.09	1.31	1.10	**2.41**	m
	isolated; not exceeding 0.50 m2	0.10	1.52	1.94	**3.46**	Nr
033	Glazed doors and screens in panes:					
	small - not exceeding 0.10 m2	0.54	8.22	1.96	**10.18**	m2
	medium - 0.10 - 0.50 m2	0.37	5.64	1.51	**7.15**	m2
	large - 0.50 - 1.00 m2	0.35	5.39	1.37	**6.76**	m2
	extra large - over 1.00 m2	0.32	4.83	1.22	**6.05**	m2
034	Windows in panes:					
	small - not exceeding 0.10 m2	0.54	8.22	1.51	**9.73**	m2
	medium - 0.10 - 0.50 m2	0.37	5.64	1.37	**7.01**	m2
	large - 0.50 - 1.00 m2	0.35	5.39	1.22	**6.61**	m2
	extra large - over 1.00 m2	0.32	4.83	1.07	**5.90**	m2
035	Structural members:					
	over 300 mm girth	0.25	3.85	3.34	**7.19**	m2
	not exceeding 150 mm girth	0.08	1.16	0.60	**1.76**	m
	150 - 300 mm girth	0.11	1.74	1.19	**2.93**	m
036	Members of roof trusses:					
	over 300 mm girth	0.32	4.87	4.50	**9.37**	m2
	not exceeding 150 mm girth	0.10	1.55	0.75	**2.30**	m
	150 - 300 mm girth	0.15	2.35	1.49	**3.84**	m
037	Radiators:					
	over 300 mm girth	0.28	4.26	4.05	**8.31**	m2
038	Pipes and conduits, ducting, trunking and the like:					
	over 300 mm girth	0.21	3.18	4.20	**7.38**	m2
039	Staircases:					
	over 300 mm girth	0.25	3.85	3.34	**7.19**	m2
	not exceeding 150 mm girth	0.08	1.16	0.60	**1.76**	m
	150 - 300 mm girth	0.11	1.74	1.19	**2.93**	m
	Note 040 not used					
041	**Prepare; one coat smooth metal finish, unprimed metalwork surfaces**					
042	General surfaces:					
	over 300 mm girth	0.19	2.89	3.01	**5.90**	m2
	not exceeding 150 mm girth	0.06	0.87	0.49	**1.36**	m

Unit Rates

		Man-Hours	Net Labour Price £	Net Mats Price £	Net Unit Price £	Unit
	150 - 300 mm girth	0.09	1.31	1.00	**2.31**	m
	isolated; not exceeding 0.50 m2	0.10	1.52	1.75	**3.27**	Nr
043	Glazed doors and screens in panes:					
	small - not exceeding 0.10 m2	0.54	8.22	1.77	**9.99**	m2
	medium - 0.10 - 0.50 m2	0.37	5.64	1.37	**7.01**	m2
	large - 0.50 - 1.00 m2	0.35	5.39	1.37	**6.76**	m2
	extra large - over 1.00 m2	0.32	4.83	1.10	**5.93**	m2
044	Windows in panes:					
	small - not exceeding 0.10 m2	0.54	8.22	1.37	**9.59**	m2
	medium - 0.10 - 0.50 m2	0.37	5.64	1.23	**6.87**	m2
	large - 0.50 - 1.00 m2	0.35	5.39	1.10	**6.49**	m2
	extra large - over 1.00 m2	0.32	4.83	0.96	**5.79**	m2
045	Structural members:					
	over 300 mm girth	0.25	3.85	3.01	**6.86**	m2
	not exceeding 150 mm girth	0.08	1.16	0.54	**1.70**	m
	150 - 300 mm girth	0.11	1.74	1.07	**2.81**	m
046	Members of roof trusses:					
	over 300 mm girth	0.32	4.87	4.05	**8.92**	m2
	not exceeding 150 mm girth	0.10	1.55	0.67	**2.22**	m
	150 - 300 mm girth	0.15	2.35	1.34	**3.69**	m
047	Radiators:					
	over 300 mm girth	0.28	4.26	3.65	**7.91**	m2
048	Pipes and conduits, ducting, trunking and the like:					
	over 300 mm girth	0.21	3.18	3.78	**6.96**	m2
049	Staircases:					
	over 300 mm girth	0.25	3.85	3.01	**6.86**	m2
	not exceeding 150 mm girth	0.08	1.16	0.54	**1.70**	m
	150 - 300 mm girth	0.11	1.74	1.07	**2.81**	m
050	**Prepare; one coat Thermoguard high build primer, primed surfaces**					
051	Structural members:					
	over 300 mm girth	0.25	3.81	1.21	**5.02**	m2
	not exceeding 150 mm girth	0.08	1.22	0.19	**1.41**	m
	150 - 300 mm girth	0.11	1.68	0.36	**2.04**	m
052	Members of roof trusses:					
	over 300 mm girth	0.28	4.19	1.21	**5.40**	m2
	not exceeding 150 mm girth	0.08	1.22	0.19	**1.41**	m
	150 - 300 mm girth	0.11	1.68	0.36	**2.04**	m

INTERNAL WORK VF

		Unit Rates	Man-Hours	Net Labour Price £	Net Mats Price £	Net Unit Price £	Unit
	053	One coat 'Thermoguard Thermacoat W' intumescent paint to high build primer, protection given 30 minutes					
	054	Structural members:					
		over 300 mm girth	0.25	3.81	1.74	**5.55**	m2
		not exceeding 150 mm girth	0.08	1.22	0.27	**1.49**	m
		150 - 300 mm girth	0.11	1.68	0.52	**2.20**	m
	055	Members of roof trusses:					
		over 300 mm girth	0.28	4.19	1.74	**5.93**	m2
		not exceeding 150 mm girth	0.09	1.34	0.27	**1.61**	m
		150 - 300 mm girth	0.12	1.84	0.52	**2.36**	m
	056	Subsequent coat of 'Thermoguard Thermacoat W' intumescent paint giving extra protection of 30 minutes					
	057	Structural members:					
		over 300 mm girth	0.18	2.74	1.74	**4.48**	m2
		not exceeding 150 mm girth	0.08	1.22	0.27	**1.49**	m
		150 - 300 mm girth	0.11	1.68	0.52	**2.20**	m
	058	Members of roof trusses:					
		over 300 mm girth	0.20	3.02	1.71	**4.73**	m2
		not exceeding 150 mm girth	0.09	1.34	0.27	**1.61**	m
		150 - 300 mm girth	0.12	1.84	0.52	**2.36**	m
	059	Prepare; second and subsequent coat 'Thermoguard flame retardant acrylic', matt/eggshell					
	060	Structural members:					
		over 300 mm girth	0.20	3.05	1.43	**4.48**	m2
		not exceeding 150 mm girth	0.09	1.37	0.22	**1.59**	m
		150 - 300 mm girth	0.13	1.90	0.42	**2.32**	m
	061	Members of roof trusses:					
		over 300 mm girth	0.22	3.35	1.47	**4.82**	m2
		not exceeding 150 mm girth	0.10	1.51	0.22	**1.73**	m
		150 - 300 mm girth	0.14	2.10	0.45	**2.55**	m
VF		**WOODWORK - INTERNALLY**					
		WOODWORK PRIMERS					
	001	**Prepare, knotting, stopping; one coat wood primer, wood surfaces**					
	002	General surfaces:					
		over 300 mm girth	0.22	3.35	1.03	**4.38**	m2
		not exceeding 150 mm girth	0.07	1.01	0.18	**1.19**	m
		150 - 300 mm girth	0.10	1.51	0.30	**1.81**	m
		isolated; not exceeding 0.50 m2	0.16	2.44	0.51	**2.95**	Nr

INTERNAL WORK

Unit Rates

		Man-Hours	Net Labour Price £	Net Mats Price £	Net Unit Price £	Unit
003	Glazed doors and screens in panes:					
	small - not exceeding 0.10 m2	0.50	7.54	0.63	**8.17**	m2
	medium - 0.10 - 0.50 m2	0.36	5.53	0.53	**6.06**	m2
	large - 0.50 - 1.00 m2	0.33	5.03	0.42	**5.45**	m2
	extra large - over 1.00 m2	0.31	4.69	0.32	**5.01**	m2
004	Windows in panes:					
	small - not exceeding 0.10 m2	0.55	8.38	0.83	**9.21**	m2
	medium - 0.10 - 0.50 m2	0.39	5.86	0.73	**6.59**	m2
	large - 0.50 - 1.00 m2	0.36	5.53	0.63	**6.16**	m2
	extra large - over 1.00 m2	0.33	5.03	0.53	**5.56**	m2
005	Frames, linings and associated mouldings:					
	over 300 mm girth	0.22	3.35	1.03	**4.38**	m2
	not exceeding 150 mm girth	0.07	1.01	0.18	**1.19**	m
	150 - 300 mm girth	0.10	1.51	0.30	**1.81**	m
006	Cornices:					
	over 300 mm girth	0.26	4.02	1.03	**5.05**	m2
	not exceeding 150 mm girth	0.08	1.17	0.18	**1.35**	m
	150 - 300 mm girth	0.12	1.84	0.30	**2.14**	m
007	Skirtings, dado rails, picture rails and the like:					
	over 300 mm girth	0.24	3.69	1.03	**4.72**	m2
	not exceeding 150 mm girth	0.07	1.11	0.18	**1.29**	m
	150 - 300 mm girth	0.11	1.68	0.30	**1.98**	m
008	Staircases:					
	over 300 mm girth	0.23	3.52	1.03	**4.55**	m2
	not exceeding 150 mm girth	0.07	1.05	0.18	**1.23**	m
	150 - 300 mm girth	0.11	1.60	0.30	**1.90**	m
009	**Prepare, knotting, stopping; one coat aluminium wood primer, wood surfaces**					
010	General surfaces:					
	over 300 mm girth	0.22	3.35	0.99	**4.34**	m2
	not exceeding 150 mm girth	0.07	1.01	0.17	**1.18**	m
	150 - 300 mm girth	0.10	1.51	0.29	**1.80**	m
	isolated; not exceeding 0.50 m2	0.16	2.44	0.49	**2.93**	Nr
011	Glazed doors and screens in panes:					
	small - not exceeding 0.10 m2	0.50	7.54	0.61	**8.15**	m2
	medium - 0.10 - 0.50 m2	0.36	5.53	0.51	**6.04**	m2
	large - 0.50 - 1.00 m2	0.33	5.03	0.41	**5.44**	m2
	extra large - over 1.00 m2	0.31	4.69	0.31	**5.00**	m2
012	Windows in panes:					
	small - not exceeding 0.10 m2	0.55	8.38	0.80	**9.18**	m2
	medium - 0.10 - 0.50 m2	0.39	5.86	0.70	**6.56**	m2
	large - 0.50 - 1.00 m2	0.36	5.53	0.61	**6.14**	m2

Unit Rates

		Man-Hours	Net Labour Price £	Net Mats Price £	Net Unit Price £	Unit
	extra large - over 1.00 m2	0.33	5.03	0.51	**5.54**	m2
013	**Frames, linings and associated mouldings:**					
	over 300 mm girth	0.22	3.35	0.99	**4.34**	m2
	not exceeding 150 mm girth	0.07	1.01	0.17	**1.18**	m
	150 - 300 mm girth	0.10	1.51	0.29	**1.80**	m
014	**Cornices:**					
	over 300 mm girth	0.26	4.02	0.99	**5.01**	m2
	not exceeding 150 mm girth	0.08	1.17	0.17	**1.34**	m
	150 - 300 mm girth	0.12	1.84	0.29	**2.13**	m
015	**Skirtings, dado rails, picture rails and the like:**					
	over 300 mm girth	0.24	3.69	0.99	**4.68**	m2
	not exceeding 150 mm girth	0.07	1.11	0.17	**1.28**	m
	150 - 300 mm girth	0.11	1.68	0.29	**1.97**	m
016	**Staircases:**					
	over 300 mm girth	0.23	3.52	0.99	**4.51**	m2
	not exceeding 150 mm girth	0.07	1.05	0.17	**1.22**	m
	150 - 300 mm girth	0.11	1.60	0.29	**1.89**	m
017	**Prepare, knotting, stopping; one coat acrylic wood primer, wood surfaces**					
018	**General surfaces:**					
	over 300 mm girth	0.22	3.35	1.56	**4.91**	m2
	not exceeding 150 mm girth	0.07	1.01	0.26	**1.27**	m
	150 - 300 mm girth	0.10	1.51	0.46	**1.97**	m
	isolated; not exceeding 0.50 m2	0.16	2.44	0.77	**3.21**	Nr
019	**Glazed doors and screens in panes:**					
	small - not exceeding 0.10 m2	0.50	7.54	0.95	**8.49**	m2
	medium - 0.10 - 0.50 m2	0.36	5.53	0.79	**6.32**	m2
	large - 0.50 - 1.00 m2	0.33	5.03	0.64	**5.67**	m2
	extra large - over 1.00 m2	0.31	4.69	0.48	**5.17**	m2
020	**Windows in panes:**					
	small - not exceeding 0.10 m2	0.55	8.38	1.25	**9.63**	m2
	medium - 0.10 - 0.50 m2	0.39	5.86	1.10	**6.96**	m2
	large - 0.50 - 1.00 m2	0.36	5.53	0.95	**6.48**	m2
	extra large - over 1.00 m2	0.33	5.03	0.79	**5.82**	m2
021	**Frames, linings and associated mouldings:**					
	over 300 mm girth	0.22	3.35	1.56	**4.91**	m2
	not exceeding 150 mm girth	0.07	1.01	0.26	**1.27**	m
	150 - 300 mm girth	0.10	1.51	0.46	**1.97**	m
022	**Cornices:**					
	over 300 mm girth	0.26	4.02	1.56	**5.58**	m2
	not exceeding 150 mm girth	0.08	1.17	0.26	**1.43**	m

Unit Rates

		Man-Hours	Net Labour Price £	Net Mats Price £	Net Unit Price £	Unit
	150 - 300 mm girth	0.12	1.84	0.46	**2.30**	m
023	Skirtings, dado rails, picture rails and the like:					
	over 300 mm girth	0.24	3.69	1.56	**5.25**	m2
	not exceeding 150 mm girth	0.07	1.11	0.26	**1.37**	m
	150 - 300 mm girth	0.11	1.68	0.46	**2.14**	m
024	Staircases:					
	over 300 mm girth	0.23	3.52	1.56	**5.08**	m2
	not exceeding 150 mm girth	0.07	1.05	0.26	**1.31**	m
	150 - 300 mm girth	0.11	1.60	0.44	**2.04**	m
	WOODWORK UNDERCOAT - INTERNALLY					
025	**One undercoat alkyd based paint, white; primed wood surfaces**					
026	General surfaces:					
	over 300 mm girth	0.17	2.51	0.52	**3.03**	m2
	not exceeding 150 mm girth	0.05	0.76	0.07	**0.83**	m
	150 - 300 mm girth	0.08	1.14	0.15	**1.29**	m
	isolated; not exceeding 0.50 m2	0.16	2.44	0.25	**2.69**	Nr
027	Glazed doors and screens in panes:					
	small - not exceeding 0.10 m2	0.44	6.70	0.32	**7.02**	m2
	medium - 0.10 - 0.50 m2	0.31	4.69	0.27	**4.96**	m2
	large - 0.50 - 1.00 m2	0.28	4.19	0.22	**4.41**	m2
	extra large - over 1.00 m2	0.25	3.85	0.17	**4.02**	m2
028	Windows in panes:					
	small - not exceeding 0.10 m2	0.50	7.54	0.42	**7.96**	m2
	medium - 0.10 - 0.50 m2	0.33	5.03	0.37	**5.40**	m2
	large - 0.50 - 1.00 m2	0.31	4.69	0.32	**5.01**	m2
	extra large - over 1.00 m2	0.28	4.19	0.27	**4.46**	m2
029	Frames, linings and associated mouldings:					
	over 300 mm girth	0.17	2.51	0.52	**3.03**	m2
	not exceeding 150 mm girth	0.05	0.76	0.07	**0.83**	m
	150 - 300 mm girth	0.08	1.14	0.15	**1.29**	m
030	Cornices:					
	over 300 mm girth	0.20	3.02	0.52	**3.54**	m2
	not exceeding 150 mm girth	0.06	0.90	0.07	**0.97**	m
	150 - 300 mm girth	0.09	1.37	0.15	**1.52**	m
031	Skirtings, dado rails, picture rails and the like:					
	over 300 mm girth	0.18	2.77	0.52	**3.29**	m2
	not exceeding 150 mm girth	0.06	0.84	0.07	**0.91**	m
	150 - 300 mm girth	0.08	1.26	0.15	**1.41**	m

		Man-Hours	Net Labour Price £	Net Mats Price £	Net Unit Price £	Unit
032	Staircases:					
	over 300 mm girth	0.17	2.65	0.52	**3.17**	m2
	not exceeding 150 mm girth	0.05	0.79	0.07	**0.86**	m
	150 - 300 mm girth	0.08	1.19	0.15	**1.34**	m
	WOODWORK ALKYD BASED GLOSS FINISH					
033	**One coat alkyd based paint gloss finish; undercoated wood surfaces**					
034	General surfaces:					
	over 300 mm girth	0.19	2.85	0.52	**3.37**	m2
	not exceeding 150 mm girth	0.06	0.85	0.07	**0.92**	m
	150 - 300 mm girth	0.09	1.29	0.15	**1.44**	m
	isolated; not exceeding 0.50 m2	0.16	2.44	0.25	**2.69**	Nr
035	Glazed doors and screens in panes:					
	small - not exceeding 0.10 m2	0.46	7.04	0.32	**7.36**	m2
	medium - 0.10 - 0.50 m2	0.33	5.03	0.27	**5.30**	m2
	large - 0.50 - 1.00 m2	0.30	4.52	0.22	**4.74**	m2
	extra large - over 1.00 m2	0.28	4.19	0.17	**4.36**	m2
036	Windows in panes:					
	small - not exceeding 0.10 m2	0.52	7.87	0.42	**8.29**	m2
	medium - 0.10 - 0.50 m2	0.35	5.36	0.37	**5.73**	m2
	large - 0.50 - 1.00 m2	0.33	5.03	0.32	**5.35**	m2
	extra large - over 1.00 m2	0.30	4.52	0.27	**4.79**	m2
037	Frames, linings and associated mouldings:					
	over 300 mm girth	0.19	2.85	0.52	**3.37**	m2
	not exceeding 150 mm girth	0.06	0.85	0.07	**0.92**	m
	150 - 300 mm girth	0.09	1.29	0.15	**1.44**	m
038	Cornices:					
	over 300 mm girth	0.22	3.35	0.52	**3.87**	m2
	not exceeding 150 mm girth	0.07	1.02	0.07	**1.09**	m
	150 - 300 mm girth	0.10	1.54	0.15	**1.69**	m
039	Skirtings, dado rails, picture rails and the like:					
	over 300 mm girth	0.21	3.14	0.52	**3.66**	m2
	not exceeding 150 mm girth	0.06	0.94	0.07	**1.01**	m
	150 - 300 mm girth	0.09	1.43	0.15	**1.58**	m
040	Staircases:					
	over 300 mm girth	0.20	3.00	0.52	**3.52**	m2
	not exceeding 150 mm girth	0.06	0.90	0.07	**0.97**	m
	150 - 300 mm girth	0.09	1.36	0.15	**1.51**	m

Unit Rates

		Man-Hours	Net Labour Price £	Net Mats Price £	Net Unit Price £	Unit
	WOODWORK, ACRYLATED RUBBER PAINT					
041	**Prepare, knotting, stopping; one coat of acrylated rubber paint primer, wood surfaces, applied to a dry film thickness of 55 microns**					
042	General surfaces:					
	over 300 mm girth	0.25	3.85	3.03	**6.88**	m2
	not exceeding 150 mm girth	0.08	1.16	0.52	**1.68**	m
	150 - 300 mm girth	0.15	2.35	0.98	**3.33**	m
	isolated; not exceeding 0.50 m2	0.16	2.44	1.64	**4.08**	Nr
043	Glazed doors and screens in panes:					
	small - not exceeding 0.10 m2	0.57	8.68	1.99	**10.67**	m2
	medium - 0.10 - 0.50 m2	0.42	6.37	1.66	**8.03**	m2
	large - 0.50 - 1.00 m2	0.38	5.79	1.37	**7.16**	m2
	extra large - over 1.00 m2	0.35	5.39	1.00	**6.39**	m2
044	Windows in panes:					
	small - not exceeding 0.10 m2	0.63	9.64	2.65	**12.29**	m2
	medium - 0.10 - 0.50 m2	0.44	6.75	2.32	**9.07**	m2
	large - 0.50 - 1.00 m2	0.42	6.37	1.99	**8.36**	m2
	extra large - over 1.00 m2	0.38	5.79	1.66	**7.45**	m2
045	Frames, linings and associated mouldings:					
	over 300 mm girth	0.25	3.85	3.24	**7.09**	m2
	not exceeding 150 mm girth	0.08	1.16	0.55	**1.71**	m
	150 - 300 mm girth	0.11	1.74	1.06	**2.80**	m
046	Cornices:					
	over 300 mm girth	0.30	4.63	3.24	**7.87**	m2
	not exceeding 150 mm girth	0.09	1.36	0.55	**1.91**	m
	150 - 300 mm girth	0.14	2.13	1.06	**3.19**	m
047	Skirtings, dado rails, picture rails and the like:					
	over 300 mm girth	0.28	4.23	3.24	**7.47**	m2
	not exceeding 150 mm girth	0.08	1.28	0.55	**1.83**	m
	150 - 300 mm girth	0.13	1.93	1.06	**2.99**	m
048	Staircases:					
	over 300 mm girth	0.27	4.05	3.03	**7.08**	m2
	not exceeding 150 mm girth	0.08	1.20	0.52	**1.72**	m
	150 - 300 mm girth	0.12	1.83	0.98	**2.81**	m
049	**One undercoat of acrylated rubber paint; primed wood surfaces; applied to a dry film thickness of 100 microns**					
050	General surfaces:					
	over 300 mm girth	0.19	2.89	9.07	**11.96**	m2
	not exceeding 150 mm girth	0.06	0.87	1.49	**2.36**	m
	150 - 300 mm girth	0.09	1.31	2.99	**4.30**	m

Unit Rates

		Man-Hours	Net Labour Price £	Net Mats Price £	Net Unit Price £	Unit
	isolated; not exceeding 0.50 m2	0.16	2.44	4.98	**7.42**	Nr
051	Glazed doors and screens in panes:					
	small - not exceeding 0.10 m2	0.51	7.71	6.00	**13.71**	m2
	medium - 0.10 - 0.50 m2	0.35	5.39	5.00	**10.39**	m2
	large - 0.50 - 1.00 m2	0.32	4.83	4.01	**8.84**	m2
	extra large - over 1.00 m2	0.29	4.45	3.01	**7.46**	m2
052	Windows in panes:					
	small - not exceeding 0.10 m2	0.54	8.22	7.99	**16.21**	m2
	medium - 0.10 - 0.50 m2	0.38	5.79	6.99	**12.78**	m2
	large - 0.50 - 1.00 m2	0.35	5.39	6.00	**11.39**	m2
	extra large - over 1.00 m2	0.32	4.83	5.00	**9.83**	m2
053	Frames, linings and associated mouldings:					
	over 300 mm girth	0.19	2.89	9.48	**12.37**	m2
	not exceeding 150 mm girth	0.06	0.87	1.49	**2.36**	m
	150 - 300 mm girth	0.09	1.31	2.99	**4.30**	m
054	Cornices:					
	over 300 mm girth	0.23	3.47	9.48	**12.95**	m2
	not exceeding 150 mm girth	0.07	1.04	1.49	**2.53**	m
	150 - 300 mm girth	0.10	1.57	2.99	**4.56**	m
055	Skirtings, dado rails, picture rails and the like:					
	over 300 mm girth	0.21	3.18	9.48	**12.66**	m2
	not exceeding 150 mm girth	0.06	0.97	1.49	**2.46**	m
	150 - 300 mm girth	0.10	1.45	2.99	**4.44**	m
056	Staircases:					
	over 300 mm girth	0.20	3.05	9.07	**12.12**	m2
	not exceeding 150 mm girth	0.06	0.90	1.49	**2.39**	m
	150 - 300 mm girth	0.09	1.37	2.99	**4.36**	m
057	**One coat of acrylated rubber paint, gloss finish; undercoated wood surfaces; applied to a dry film thickness of 50 microns**					
058	General surfaces:					
	over 300 mm girth	0.22	3.29	1.90	**5.19**	m2
	not exceeding 150 mm girth	0.07	0.99	0.31	**1.30**	m
	150 - 300 mm girth	0.10	1.49	0.62	**2.11**	m
	isolated; not exceeding 0.50 m2	0.16	2.44	1.03	**3.47**	Nr
059	Glazed doors and screens in panes:					
	small - not exceeding 0.10 m2	0.53	8.09	1.26	**9.35**	m2
	medium - 0.10 - 0.50 m2	0.38	5.79	1.06	**6.85**	m2
	large - 0.50 - 1.00 m2	0.34	5.21	0.85	**6.06**	m2
	extra large - over 1.00 m2	0.32	4.83	0.68	**5.51**	m2

Unit Rates — INTERNAL WORK VF

		Man-Hours	Net Labour Price £	Net Mats Price £	Net Unit Price £	Unit
060	Windows in panes:					
	small - not exceeding 0.10 m2	0.60	9.06	1.67	**10.73**	m2
	medium - 0.10 - 0.50 m2	0.43	6.47	1.47	**7.94**	m2
	large - 0.50 - 1.00 m2	0.38	5.79	1.26	**7.05**	m2
	extra large - over 1.00 m2	0.34	5.21	1.06	**6.27**	m2
061	Frames, linings and associated mouldings:					
	over 300 mm girth	0.22	3.29	2.23	**5.52**	m2
	not exceeding 150 mm girth	0.07	0.99	0.37	**1.36**	m
	150 - 300 mm girth	0.10	1.49	0.73	**2.22**	m
062	Cornices:					
	over 300 mm girth	0.25	3.85	2.23	**6.08**	m2
	not exceeding 150 mm girth	0.08	1.17	0.37	**1.54**	m
	150 - 300 mm girth	0.12	1.78	0.73	**2.51**	m
063	Skirtings, dado rails, picture rails and the like:					
	over 300 mm girth	0.24	3.61	2.23	**5.84**	m2
	not exceeding 150 mm girth	0.07	1.07	0.37	**1.44**	m
	150 - 300 mm girth	0.11	1.64	0.73	**2.37**	m
064	Staircases:					
	over 300 mm girth	0.23	3.46	1.90	**5.36**	m2
	not exceeding 150 mm girth	0.07	1.04	0.28	**1.32**	m
	150 - 300 mm girth	0.10	1.55	0.62	**2.17**	m
	WOODWORK INTUMESCENT PAINT					
065	**One coat and subsequent coats Thermoguard Timber Coat to primed**					
066	General surfaces:					
	over 300 mm girth	0.19	2.89	2.84	**5.73**	m2
	not exceeding 150 mm girth	0.06	0.91	0.42	**1.33**	m
	150 - 300 mm girth	0.09	1.37	0.85	**2.22**	m
	isolated; not exceeding 0.5 m2	0.11	1.74	1.27	**3.01**	nr
067	Fire resistant glazed doors, screens and windows:					
	small - not exceeding 0.10 m2	0.51	7.77	1.91	**9.68**	m2
	medium - 0.10 - 0.50 m2	0.38	5.79	1.58	**7.37**	m2
	large - 0.50 - 1 m2	0.33	5.03	1.28	**6.31**	m2
	extra large - over 1 m2	0.28	4.19	1.15	**5.34**	m2
	opening edge	0.07	1.07	0.30	**1.37**	m
068	Staircases:					
	over 300 mm girth	0.21	3.18	3.12	**6.30**	m2
	not exceeding 150 mm girth	0.07	1.01	0.46	**1.47**	m
	150 - 300 mm girth	0.10	1.51	0.93	**2.44**	m
	isolated; not exceeding 0.5 m2	0.13	1.90	1.39	**3.29**	nr
	balustrade (measured both sides)	0.16	2.47	17.51	**19.98**	m2

Unit Rates

		Man-Hours	Net Labour Price £	Net Mats Price £	Net Unit Price £	Unit
	Note 069 - 100 not used					
	WOODWORK, STAINING					
101	**Prepare; one coat interior satin sheen wood stain**					
102	General surfaces:					
	over 300 mm girth	0.13	1.98	1.45	**3.43**	m2
	not exceeding 150 mm girth	0.03	0.46	2.85	**3.31**	m
	150 - 300 mm girth	0.06	0.91	2.85	**3.76**	m
	isolated; not exceeding 0.50 m2	0.07	1.07	0.71	**1.78**	Nr
103	Glazed doors and screens in panes:					
	small - not exceeding 0.10 m2	0.28	4.26	1.02	**5.28**	m2
	medium - 0.10 - 0.50 m2	0.21	3.20	0.88	**4.08**	m2
	large - 0.50 - 1.00 m2	0.20	3.05	0.74	**3.79**	m2
	extra large - over 1.00 m2	0.19	2.89	0.59	**3.48**	m2
104	Windows in panes:					
	small - not exceeding 0.10 m2	0.31	4.72	1.02	**5.74**	m2
	medium - 0.10 - 0.50 m2	0.23	3.50	0.88	**4.38**	m2
	large - 0.50 - 1.00 m2	0.22	3.35	0.74	**4.09**	m2
	extra large - over 1.00 m2	0.21	3.20	0.59	**3.79**	m2
105	Frames, linings and associated mouldings:					
	over 300 mm girth	0.13	1.98	1.45	**3.43**	m2
	not exceeding 150 mm girth	0.03	0.46	2.85	**3.31**	m
	150 - 300 mm girth	0.06	0.91	0.29	**1.20**	m
106	Skirtings, dado rails, picture rails and the like:					
	over 300 mm girth	0.13	1.98	1.45	**3.43**	m2
	not exceeding 150 mm girth	0.03	0.46	0.29	**0.75**	m
	150 - 300 mm girth	0.06	0.91	0.29	**1.20**	m
107	Staircases:					
	over 300 mm girth	0.13	1.98	1.45	**3.43**	m2
	not exceeding 150 mm girth	0.03	0.46	0.29	**0.75**	m
	150 - 300 mm girth	0.06	0.91	0.29	**1.20**	m
VG	**CLEAR FINISHES ON WOODWORK - INTERNALLY**					
	WOODWORK POLYURETHANE - INTERNALLY					
001	**Prepare; one coat polyurethane varnish, first coat on unprimed wood surfaces**					
002	General surfaces:					
	over 300 mm girth	0.22	3.35	1.03	**4.38**	m2
	not exceeding 150 mm girth	0.07	1.01	0.19	**1.20**	m
	150 - 300 mm girth	0.10	1.51	0.36	**1.87**	m
	isolated; not exceeding 0.50 m2	0.16	2.44	0.53	**2.97**	Nr

Unit Rates

INTERNAL WORK VG

		Man-Hours	Net Labour Price £	Net Mats Price £	Net Unit Price £	Unit
003	Glazed doors and screens in panes:					
	small - not exceeding 0.10 m2	0.58	8.83	0.64	**9.47**	m2
	medium - 0.10 - 0.50 m2	0.36	5.53	0.55	**6.08**	m2
	large - 0.50 - 1.00 m2	0.33	5.03	0.45	**5.48**	m2
	extra large - over 1.00 m2	0.31	4.69	0.34	**5.03**	m2
004	Windows in panes:					
	small - not exceeding 0.10 m2	0.63	9.59	0.84	**10.43**	m2
	medium - 0.10 - 0.50 m2	0.39	5.86	0.74	**6.60**	m2
	large - 0.50 - 1.00 m2	0.36	5.53	0.64	**6.17**	m2
	extra large - over 1.00 m2	0.33	5.03	0.55	**5.58**	m2
005	Frames, linings and associated mouldings:					
	over 300 mm girth	0.22	3.35	1.11	**4.46**	m2
	not exceeding 150 mm girth	0.07	1.01	0.21	**1.22**	m
	150 - 300 mm girth	0.10	1.51	0.39	**1.90**	m
006	Cornices:					
	over 300 mm girth	0.26	4.02	1.11	**5.13**	m2
	not exceeding 150 mm girth	0.08	1.17	0.21	**1.38**	m
	150 - 300 mm girth	0.12	1.84	0.39	**2.23**	m
007	Skirtings, dado rails, picture rails and the like:					
	over 300 mm girth	0.24	3.69	1.11	**4.80**	m2
	not exceeding 150 mm girth	0.07	1.11	0.22	**1.33**	m
	150 - 300 mm girth	0.11	1.68	0.39	**2.07**	m
008	Staircases:					
	over 300 mm girth	0.23	3.52	1.03	**4.55**	m2
	not exceeding 150 mm girth	0.07	1.05	0.19	**1.24**	m
	150 - 300 mm girth	0.11	1.60	0.36	**1.96**	m
009	Floor surfaces	0.06	0.84	1.03	**1.87**	m2
010	**Prepare; one coat polyurethane varnish, second and subsequent coats**					
011	General surfaces:					
	over 300 mm girth	0.19	2.85	0.80	**3.65**	m2
	not exceeding 150 mm girth	0.06	0.85	0.13	**0.98**	m
	150 - 300 mm girth	0.09	1.29	0.26	**1.55**	m
	isolated; not exceeding 0.50 m2	0.16	2.44	0.39	**2.83**	Nr
012	Glazed doors and screens in panes:					
	small - not exceeding 0.10	0.55	8.38	0.49	**8.87**	m2
	medium - 0.10 - 0.50 m2	0.33	5.03	0.41	**5.44**	m2
	large - 0.50 - 1.00 m2	0.30	4.52	0.34	**4.86**	m2
	extra large - over 1.00 m2	0.28	4.19	0.26	**4.45**	m2
013	Windows in panes:					
	small - not exceeding 0.10 m2	0.59	8.99	0.65	**9.64**	m2

Unit Rates

		Man-Hours	Net Labour Price £	Net Mats Price £	Net Unit Price £	Unit
	medium - 0.10 - 0.50 m2	0.35	5.36	0.57	**5.93**	m2
	large - 0.50 - 1.00 m2	0.33	5.03	0.49	**5.52**	m2
	extra large - over 1.00 m2	0.30	4.52	0.41	**4.93**	m2
014	Frames, linings and associated mouldings:					
	over 300 mm girth	0.19	2.85	0.86	**3.71**	m2
	not exceeding 150 mm girth	0.06	0.85	0.14	**0.99**	m
	150 - 300 mm girth	0.09	1.29	0.27	**1.56**	m
015	Cornices:					
	over 300 mm girth	0.22	3.35	0.86	**4.21**	m2
	not exceeding 150 mm girth	0.07	1.02	0.14	**1.16**	m
	150 - 300 mm girth	0.10	1.54	0.27	**1.81**	m
016	Skirtings, dado rails, picture rails and the like:					
	over 300 mm girth	0.21	3.14	0.86	**4.00**	m2
	not exceeding 150 mm girth	0.06	0.94	0.14	**1.08**	m
	150 - 300 mm girth	0.09	1.43	0.27	**1.70**	m
017	Staircases:					
	over 300 mm girth	0.20	3.00	0.80	**3.80**	m2
	not exceeding 150 mm girth	0.06	0.90	0.13	**1.03**	m
	150 - 300 mm girth	0.09	1.36	0.26	**1.62**	m
018	Floor surfaces	0.04	0.67	0.80	**1.47**	m2

CLEAR FINISHES ON WOODWORK - INTERNALLY

		Man-Hours	Net Labour Price £	Net Mats Price £	Net Unit Price £	Unit
019	**Prepare; one coat Thermoguard fire varnish, unprimed wood surfaces**					
020	General surfaces:					
	over 300 mm girth	0.22	3.35	2.58	**5.93**	m2
	not exceeding 150 mm girth	0.07	1.05	0.39	**1.44**	m
	150 - 300 mm girth	0.10	1.58	0.77	**2.35**	m
	isolated; not exceeding 0.5 m2	0.13	2.01	1.17	**3.18**	nr
021	Fire resistant glazed doors, screens and windows:					
	small - not exceeding 0.10 m2	0.54	8.22	1.74	**9.96**	m2
	medium - 0.10 - 0.50 m2	0.38	5.79	1.43	**7.22**	m2
	large - 0.50 - 1 m2	0.33	5.03	1.17	**6.20**	m2
	extra large - over 1 m2	0.31	4.72	10.29	**15.01**	m2
	opening edge	0.07	1.07	0.28	**1.35**	m
022	Staircases:					
	over 300 mm girth	0.23	3.50	2.58	**6.08**	m2
	not exceeding 150 mm girth	0.07	1.07	0.39	**1.46**	m
	150 - 300 mm girth	0.11	1.68	0.77	**2.45**	m
	isolated; not exceeding 0.5 m2	0.14	2.13	1.16	**3.29**	m
	balustrade (measured both sides)	0.19	2.89	2.20	**5.09**	m2

INTERNAL WORK

	Unit Rates	Man-Hours	Net Labour Price £	Net Mats Price £	Net Unit Price £	Unit
023	**Prepare; one coat Thermoguard fire varnish, second and subsequent coats**					
024	General surfaces:					
	over 300 mm girth	0.19	2.89	1.56	**4.45**	m2
	not exceeding 150 mm girth	0.06	0.91	0.23	**1.14**	m
	150 - 300 mm girth	0.09	1.37	0.47	**1.84**	m
	isolated; not exceeding 0.5 m2	0.11	1.74	0.70	**2.44**	m
025	Fire resistant glazed doors, screens and windows:					
	small - not exceeding 0.10 m2	0.51	7.77	1.05	**8.82**	m2
	medium - 0.10 - 0.50 m2	0.38	5.79	0.87	**6.66**	m2
	large - 0.50 - 1 m2	0.33	5.03	0.71	**5.74**	m2
	extra large - over 1 m2	0.28	4.19	0.64	**4.83**	m2
	opening edge	0.07	1.07	0.16	**1.23**	m
026	Staircases:					
	over 300 mm girth	0.20	3.05	0.18	**3.23**	m2
	not exceeding 150 mm girth	0.07	0.99	0.23	**1.22**	m
	150 - 300 mm girth	0.10	1.52	0.47	**1.99**	m
	isolated; not exceeding 0.5 m2	0.13	1.98	0.70	**2.68**	m
	balustrade (measured both sides)	0.17	2.59	1.33	**3.92**	m2

NEW WORK EXTERNALLY

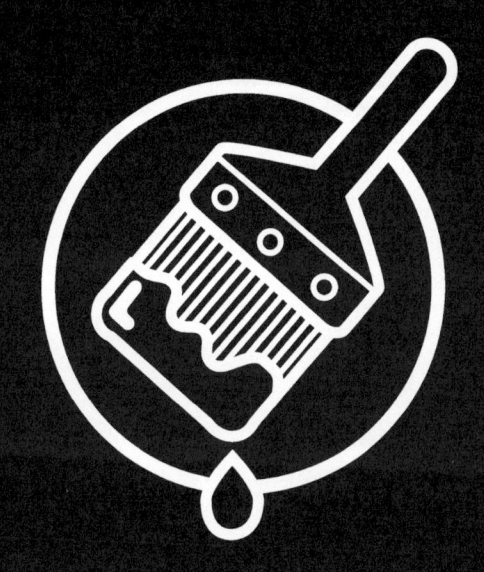

BCIS 50 years celebrating excellence

CONSTRUCTION

LESS DESK TIME MORE FREE TIME

THE REVOLUTION OF THE PRICE BOOK IS HERE
BCIS ONLINE RATES DATABASE

As a purchaser of the 2012 price book, we would like to offer you a free two week trial of our Online Rates Database. It's the price book but online, with lots of additional features to help you to:

- locate prices quickly using faster navigation
- adjust your data to suit a time frame of your choice and choose location factors to make your costs more accurate.

Plus:

- everything you need is in one place – you have a full library at your disposal
- with the full service you can download data into an Excel spreadsheet and manipulate and store your data electronically*.

*During the two week trial data downloads are not available

Offering immediate online access to independent BCIS resource rates data, quantity surveyors and others in the construction industry can have all the information needed to compile and check estimates on their desktops. You won't need to worry about being able to lay your hands on the office copy of the latest price books, all of the information is now easily accessible online.

For a **FREE TRIAL** of BCIS online rates database, register at **www.bcis.co.uk/ordbdemo**

- Accuracy
- Futureproof
- Value for money
- Saves time

- Flexible
- Customise
- Portable
- Comprehensive

BCIS is the Building Cost Information Service of **RICS** the mark of property professionalism worldwide

EXTERNAL WORK

NEW WORK EXTERNALLY

Prices in this section are given separately for first or priming coats (which include for preparation of the surfaces to be painted), undercoats and finishing coats. The total cost of decoration will be combinations of these individual coat prices in accordance with the required specification.

A large number of permutations for various specifications is therefore possible.

Refer to the Composite Prices section for examples of prices for composite painting and decorating items.

EXTERNAL WORK VH

Unit Rates

		Man-Hours	Net Labour Price £	Net Mats Price £	Net Unit Price £	Unit
VH	**NEW WORK EXTERNALLY**					
	SEE ALSO SECTION VZ FOR COMPOSITE EXAMPLES					
	WALLS - EXTERNALLY					
	MASONRY SEALER - WALLS EXTERNALLY					
001	**One coat of masonry sealer to unprimed surfaces**					
002	To walls:					
	smooth concrete	0.17	2.51	1.33	**3.84**	m2
	cement rendered	0.15	2.30	1.41	**3.71**	m2
	fair face brickwork	0.20	3.00	1.54	**4.54**	m2
	fair face blockwork	0.23	3.52	1.85	**5.37**	m2
	rough cast/pebbledash rendered	0.29	4.40	3.05	**7.45**	m2
	Tyrolean rendered	0.39	5.86	6.08	**11.94**	m2
	PRIMER SEALER - WALLS EXTERNALLY					
003	**One coat of acrylic primer/undercoat, to unprimed surfaces**					
004	To walls:					
	smooth concrete	0.22	3.35	1.05	**4.40**	m2
	cement rendered	0.19	2.94	1.12	**4.06**	m2
	fair face brickwork	0.26	3.90	1.22	**5.12**	m2
	fair face blockwork	0.31	4.69	1.47	**6.16**	m2
	rough cast/pebbledash rendered	0.39	5.86	2.41	**8.27**	m2
	Tyrolean rendered	0.51	7.81	4.81	**12.62**	m2
	ALKALI RESISTING PRIMER - WALLS EXTERNALLY					
005	**One coat of alkali resisting primer to unprimed surfaces**					
006	To walls:					
	smooth concrete	0.22	3.35	1.42	**4.77**	m2
	fibre cement	0.22	3.35	1.40	**4.75**	m2
	cement rendered	0.19	2.94	1.81	**4.75**	m2
	fair face brickwork	0.26	3.90	1.82	**5.72**	m2
	fair face blockwork	0.31	4.69	2.54	**7.23**	m2
	rough cast/pebbledash rendered	0.39	5.86	4.20	**10.06**	m2
	Tyrolean rendered	0.51	7.81	8.39	**16.20**	m2
	EMULSION PAINT, MATT - WALLS EXTERNALLY					
007	**One coat of emulsion paint, matt finish (white); first coat to unprimed surfaces**					
008	To walls:					
	smooth concrete	0.17	2.51	0.46	**2.97**	m2
	fibre cement	0.17	2.51	0.45	**2.96**	m2

Unit Rates

EXTERNAL WORK

	Man-Hours	Net Labour Price £	Net Mats Price £	Net Unit Price £	Unit
cement rendered	0.15	2.30	0.54	**2.84**	m2
fair face brickwork	0.20	3.00	0.59	**3.59**	m2
fair face blockwork	0.23	3.52	0.71	**4.23**	m2
rough cast/pebbledash rendered	0.29	4.40	1.60	**6.00**	m2
Tyrolean rendered	0.39	5.86	3.18	**9.04**	m2

009 One coat of emulsion paint, matt finish (white); second and subsequent coats

010 To walls:

	Man-Hours	Net Labour	Net Mats	Net Unit	Unit
smooth concrete	0.12	1.75	0.42	**2.17**	m2
fibre cement	0.12	1.75	0.41	**2.16**	m2
cement rendered	0.11	1.60	0.45	**2.05**	m2
fair face brickwork	0.14	2.19	0.50	**2.69**	m2
fair face blockwork	0.16	2.47	0.56	**3.03**	m2
rough cast/pebbledash rendered	0.23	3.52	1.21	**4.73**	m2
Tyrolean rendered	0.29	4.40	2.40	**6.80**	m2

MASONRY PAINT - WALLS EXTERNALLY

011 One coat of Masonry paint on sealed surfaces

012 To walls:

	Man-Hours	Net Labour	Net Mats	Net Unit	Unit
smooth concrete	0.12	1.75	0.69	**2.44**	m2
fibre cement	0.12	1.75	0.68	**2.43**	m2
cement rendered	0.11	1.60	0.79	**2.39**	m2
fair face brickwork	0.14	2.19	0.80	**2.99**	m2
fair face blockwork	0.16	2.47	0.96	**3.43**	m2
rough cast/pebbledash rendered	0.23	3.52	1.18	**4.70**	m2
Tyrolean rendered	0.29	4.40	1.89	**6.29**	m2

013 One coat of Masonry paint; first coat to unprimed surfaces

014 To walls:

	Man-Hours	Net Labour	Net Mats	Net Unit	Unit
smooth concrete	0.17	2.59	0.77	**3.36**	m2
fibre cement	0.15	2.28	0.76	**3.04**	m2
cement rendered	0.15	2.28	0.90	**3.18**	m2
fair face brickwork	0.20	3.05	0.91	**3.96**	m2
fair face blockwork	0.23	3.50	1.06	**4.56**	m2
rough cast/pebbledash rendered	0.29	4.42	1.64	**6.06**	m2
Tyrolean rendered	0.38	5.79	3.23	**9.02**	m2

Note 015 - 022 not used

SNOWCEM PAINT - WALLS EXTERNALLY

023 One coat of Snowcem cement paint; first coat including base coat of stabilising solution to unprimed surfaces

024 To walls:

	Man-Hours	Net Labour	Net Mats	Net Unit	Unit
smooth concrete	0.28	4.28	0.51	**4.79**	m2

Unit Rates

		Man-Hours	Net Labour Price £	Net Mats Price £	Net Unit Price £	Unit
	fibre cement	0.28	4.28	0.49	**4.77**	m2
	cement rendered	0.25	3.87	0.58	**4.45**	m2
	fair face brickwork	0.34	5.19	0.60	**5.79**	m2
	fair face blockwork	0.39	5.99	0.74	**6.73**	m2
	rough cast/pebbledash rendered	0.52	7.92	1.06	**8.98**	m2
	Tyrolean rendered	0.67	10.25	2.12	**12.37**	m2
025	**One coat of Snowcem cement paint; second and subsequent coats**					
026	To walls:					
	smooth concrete	0.12	1.75	0.23	**1.98**	m2
	fibre cement	0.12	1.75	0.22	**1.97**	m2
	cement rendered	0.11	1.60	0.25	**1.85**	m2
	fair face brickwork	0.14	2.19	0.26	**2.45**	m2
	fair face blockwork	0.16	2.47	0.31	**2.78**	m2
	rough cast/pebbledash rendered	0.23	3.52	0.37	**3.89**	m2
	Tyrolean rendered	0.29	4.40	0.73	**5.13**	m2
	SANDTEX MATT - WALLS EXTERNALLY					
027	**One coat of Sandtex Matt including base coat of stabilising solution to unprimed surfaces**					
028	To walls:					
	smooth concrete	0.28	4.28	1.53	**5.81**	m2
	fibre cement	0.28	4.28	1.51	**5.79**	m2
	cement rendered	0.25	3.87	1.76	**5.63**	m2
	fair face brickwork	0.34	5.19	1.79	**6.98**	m2
	fair face blockwork	0.39	5.99	1.50	**7.49**	m2
	rough cast	0.50	7.57	2.63	**10.20**	m2
	Tyrolean spardash	0.62	9.49	5.66	**15.15**	m2
029	**One coat of Sandtex Fine Build textured finish**					
030	To walls:					
	smooth concrete	0.22	3.35	3.55	**6.90**	m2
	fibre cement	0.22	3.35	3.54	**6.89**	m2
	cement rendered	0.21	3.18	4.41	**7.59**	m2
	fair face brickwork	0.28	4.19	5.30	**9.49**	m2
	fair face blockwork	0.30	4.52	5.31	**9.83**	m2
	rough cast	0.38	5.73	6.62	**12.35**	m2
	Tyrolean spardash	0.47	7.17	7.08	**14.25**	m2

EXTERNAL WORK VJ

Unit Rates

		Man-Hours	Net Labour Price £	Net Mats Price £	Net Unit Price £	Unit
VJ	**WALLS, SPRAYED FINISHES - EXTERNALLY**					
	EMULSION PAINT, MATT SPRAYED - EXTERNALLY					
	Note 001 not used					
002	**One coat of emulsion paint (sprayed), matt finish (white); first coat to unprimed surfaces**					
003	Walls over 300 mm wide:					
	smooth concrete	0.05	1.10	0.57	**1.67**	m2
	fibre cement	0.05	1.10	0.69	**1.79**	m2
	cement rendered	0.05	1.26	0.57	**1.83**	m2
	fair face brickwork	0.06	1.41	0.78	**2.19**	m2
	fair face blockwork	0.06	1.48	0.78	**2.26**	m2
	rough cast/pebbledash rendered	0.07	1.62	1.17	**2.79**	m2
	Tyrolean rendered	0.09	2.07	4.69	**6.76**	m2
004	**One coat of emulsion paint (sprayed), matt finish (white); second and subsequent coats**					
005	Walls over 300 mm wide:					
	smooth concrete	0.04	1.00	0.53	**1.53**	m2
	fibre cement	0.04	1.00	0.52	**1.52**	m2
	cement rendered	0.05	1.12	0.65	**1.77**	m2
	fair face brickwork	0.06	1.31	0.74	**2.05**	m2
	fair face blockwork	0.06	1.43	0.74	**2.17**	m2
	rough cast/pebbledash rendered	0.06	1.50	0.80	**2.30**	m2
	Tyrolean rendered	0.08	1.88	2.35	**4.23**	m2
	MASONRY PAINT - SPRAYED FINISHES					
006	**One coat of masonry paint (sprayed), on sealed surfaces**					
007	To walls:					
	smooth concrete	0.04	1.00	0.93	**1.93**	m2
	fibre cement	0.04	1.00	1.14	**2.14**	m2
	cement rendered	0.05	1.12	0.93	**2.05**	m2
	fair face brickwork	0.06	1.31	1.28	**2.59**	m2
	fair face blockwork	0.06	1.43	1.28	**2.71**	m2
	rough cast/pebbledash rendered	0.06	1.50	1.92	**3.42**	m2
	Tyrolean rendered	0.08	1.88	7.66	**9.54**	m2
VK	**METALWORK - EXTERNALLY**					
	METALWORK PRIMERS - EXTERNALLY					
001	**Prepare; one coat etching primer, metalwork surfaces, applied to a dry film thickness of 3 microns**					
002	General surfaces:					
	over 300 mm girth	0.23	3.52	0.59	**4.11**	m2

	Unit Rates	Man-Hours	Net Labour Price £	Net Mats Price £	Net Unit Price £	Unit
	not exceeding 150 mm girth	0.07	1.05	0.10	**1.15**	m
	150 - 300 mm girth	0.11	1.60	0.21	**1.81**	m
	isolated; not exceeding 0.50 m2	0.16	2.44	0.31	**2.75**	Nr
003	Glazed doors and screens in panes:					
	small - not exceeding 0.10 m2	0.58	8.80	0.33	**9.13**	m2
	medium - 0.10 - 0.50 m2	0.41	6.17	0.28	**6.45**	m2
	large - 0.50 - 1.00 m2	0.38	5.82	0.23	**6.05**	m2
	extra large - over 1.00 m2	0.35	5.28	0.23	**5.51**	m2
004	Windows in panes:					
	small - not exceeding 0.10 m2	0.58	8.80	0.28	**9.08**	m2
	medium - 0.10 - 0.50 m2	0.41	6.17	0.23	**6.40**	m2
	large - 0.50 - 1.00 m2	0.38	5.82	0.23	**6.05**	m2
	extra large - over 1.00 m2	0.35	5.28	0.18	**5.46**	m2
005	Edges of opening casements	0.06	0.88	0.05	**0.93**	m
006	Structural members:					
	over 300 mm girth	0.27	4.05	0.59	**4.64**	m2
	not exceeding 150 mm girth	0.08	1.20	0.10	**1.30**	m
	150 - 300 mm girth	0.12	1.83	0.21	**2.04**	m
007	Each side of ornamental railings, gates and the like (grouped together) measured both sides overall regardless of voids:					
	over 300 mm girth	0.18	2.80	1.31	**4.11**	m2
008	Pipes, conduits, ducting, trunking and the like:					
	over 300 mm girth	0.25	3.87	0.69	**4.56**	m2
	not exceeding 150 mm girth	0.08	1.16	0.10	**1.26**	m
	150 - 300 mm girth	0.11	1.74	0.21	**1.95**	m
009	Eaves gutters:					
	over 300 mm girth	0.25	3.87	0.69	**4.56**	m2
	not exceeding 150 mm girth	0.08	1.16	0.10	**1.26**	m
	150 - 300 mm girth	0.11	1.74	0.21	**1.95**	m
010	Staircases:					
	over 300 mm girth	0.25	3.87	0.59	**4.46**	m2
	not exceeding 150 mm girth	0.08	1.16	0.10	**1.26**	m
	150 - 300 mm girth	0.11	1.74	0.21	**1.95**	m
011	**Prepare; one coat zinc phosphate primer, metalwork surfaces**					
012	General surfaces:					
	over 300 mm girth	0.23	3.52	0.95	**4.47**	m2
	not exceeding 150 mm girth	0.07	1.05	0.14	**1.19**	m
	150 - 300 mm girth	0.11	1.60	0.28	**1.88**	m
	isolated; not exceeding 0.50 m2	0.16	2.44	0.46	**2.90**	Nr

Unit Rates

		Man-Hours	Net Labour Price £	Net Mats Price £	Net Unit Price £	Unit
013	Glazed doors and screens in panes:					
	small - not exceeding 0.10 m2	0.58	8.80	0.49	**9.29**	m2
	medium - 0.10 - 0.50 m2	0.41	6.17	0.39	**6.56**	m2
	large - 0.50 - 1.00 m2	0.38	5.82	0.34	**6.16**	m2
	extra large - over 1.00 m2	0.35	5.28	0.30	**5.58**	m2
014	Windows in panes:					
	small - not exceeding 0.10 m2	0.59	8.99	0.39	**9.38**	m2
	medium - 0.10 - 0.50 m2	0.41	6.17	0.34	**6.51**	m2
	large - 0.50 - 1.00 m2	0.38	5.82	0.30	**6.12**	m2
	extra large - over 1.00 m2	0.35	5.28	0.25	**5.53**	m2
015	Edges of opening casements	0.06	0.88	0.07	**0.95**	m
016	Structural members:					
	over 300 mm girth	0.27	4.05	0.95	**5.00**	m2
	not exceeding 150 mm girth	0.08	1.20	0.14	**1.34**	m
	150 - 300 mm girth	0.12	1.83	0.28	**2.11**	m
017	Each side of ornamental railings, gates and the like (grouped together) measured both sides overall regardless of voids:					
	over 300 mm girth	0.18	2.80	0.65	**3.45**	m2
018	Pipes and conduits, ducting, trunking and the like:					
	over 300 mm girth	0.25	3.87	1.00	**4.87**	m2
	not exceeding 150 mm girth	0.08	1.16	0.15	**1.31**	m
	150 - 300 mm girth	0.11	1.74	0.29	**2.03**	m
019	Eaves gutters:					
	over 300 mm girth	0.25	3.87	1.00	**4.87**	m2
	not exceeding 150 mm girth	0.08	1.16	0.15	**1.31**	m
	150 - 300 mm girth	0.11	1.74	0.29	**2.03**	m
020	Staircases:					
	over 300 mm girth	0.25	3.87	0.95	**4.82**	m2
	not exceeding 150 mm girth	0.08	1.16	0.14	**1.30**	m
	150 - 300 mm girth	0.11	1.74	0.28	**2.02**	m
021	**Prepare; one coat zinc chromate primer, metalwork surfaces**					
022	General surfaces:					
	over 300 mm girth	0.23	3.52	0.92	**4.44**	m2
	not exceeding 150 mm girth	0.07	1.05	0.13	**1.18**	m
	150 - 300 mm girth	0.11	1.60	0.27	**1.87**	m
	isolated; not exceeding 0.50 m2	0.16	2.44	0.45	**2.89**	Nr
023	Glazed doors and screens in panes:					
	small - not exceeding 0.10 m2	0.58	8.80	0.47	**9.27**	m2
	medium - 0.10 - 0.50 m2	0.41	6.17	0.38	**6.55**	m2
	large - 0.50 - 1.00 m2	0.38	5.82	0.33	**6.15**	m2

EXTERNAL WORK VK

	Unit Rates	Man-Hours	Net Labour Price £	Net Mats Price £	Net Unit Price £	Unit
	extra large - over 1.00 m2	0.35	5.28	0.29	**5.57**	m2
024	Windows in panes:					
	small - not exceeding 0.10 m2	0.58	8.80	0.38	**9.18**	m2
	medium - 0.10 - 0.50 m2	0.41	6.17	0.33	**6.50**	m2
	large - 0.50 - 1.00 m2	0.38	5.82	0.29	**6.11**	m2
	extra large - over 1.00 m2	0.35	5.28	0.25	**5.53**	m2
025	Edges of opening casements	0.06	0.88	0.07	**0.95**	m
026	Structural members:					
	over 300 mm girth	0.27	4.05	0.92	**4.97**	m2
	not exceeding 150 mm girth	0.08	1.20	0.13	**1.33**	m
	150 - 300 mm girth	0.12	1.83	0.27	**2.10**	m
027	Each side of ornamental railings, gates and the like (grouped together) measured both sides overall regardless of voids:					
	over 300 mm girth	0.18	2.80	0.63	**3.43**	m2
028	Pipes and conduits, ducting, trunking and the like:					
	over 300 mm girth	0.25	3.87	0.97	**4.84**	m2
	not exceeding 150 mm girth	0.08	1.16	0.14	**1.30**	m
	150 - 300 mm girth	0.11	1.74	0.29	**2.03**	m
029	Eaves gutters:					
	over 300 mm girth	0.25	3.87	0.97	**4.84**	m2
	not exceeding 150 mm girth	0.08	1.16	0.14	**1.30**	m
	150 - 300 mm girth	0.11	1.74	0.29	**2.03**	m
030	Staircases:					
	over 300 mm girth	0.25	3.87	0.92	**4.79**	m2
	not exceeding 150 mm girth	0.08	1.16	0.13	**1.29**	m
	150 - 300 mm girth	0.11	1.74	0.27	**2.01**	m
031	**Prepare; one coat zinc rich primer, metalwork surfaces**					
032	General surfaces:					
	over 300 mm girth	0.23	3.52	4.54	**8.06**	m2
	not exceeding 150 mm girth	0.07	1.05	0.67	**1.72**	m
	150 - 300 mm girth	0.11	1.60	1.36	**2.96**	m
	isolated; not exceeding 0.50 m2	0.16	2.44	2.27	**4.71**	Nr
033	Glazed doors and screens in panes:					
	small - not exceeding 0.10 m2	0.58	8.80	2.29	**11.09**	m2
	medium - 0.10 - 0.50 m2	0.41	6.17	1.83	**8.00**	m2
	large - 0.50 - 1.00 m2	0.38	5.82	1.60	**7.42**	m2
	extra large - over 1.00 m2	0.35	5.28	1.39	**6.67**	m2
034	Windows in panes:					
	small - not exceeding 0.10 m2	0.58	8.80	1.83	**10.63**	m2

		Man-Hours	Net Labour Price £	Net Mats Price £	Net Unit Price £	Unit
	medium - 0.10 - 0.50 m2	0.41	6.17	1.60	**7.77**	m2
	large - 0.50 - 1.00 m2	0.38	5.82	1.39	**7.21**	m2
	extra large - over 1.00 m2	0.35	5.28	1.16	**6.44**	m2
035	Edges of opening casements	0.06	0.88	0.35	**1.23**	m
036	Structural members:					
	over 300 mm girth	0.27	4.05	4.54	**8.59**	m2
	not exceeding 150 mm girth	0.08	1.20	0.67	**1.87**	m
	150 - 300 mm girth	0.12	1.83	1.36	**3.19**	m
037	Pipes and conduits, ducting, trunking and the like:					
	over 300 mm girth	0.25	3.87	4.83	**8.70**	m2
	not exceeding 150 mm girth	0.08	1.16	0.73	**1.89**	m
	150 - 300 mm girth	0.11	1.74	1.44	**3.18**	m
038	Each side of ornamental railings, gates and the like (grouped together) measured both sides overall regardless of voids:					
	over 300 mm girth	0.18	2.80	3.10	**5.90**	m2
039	Eaves gutters:					
	over 300 mm girth	0.25	3.87	4.83	**8.70**	m2
	not exceeding 150 mm girth	0.08	1.16	0.73	**1.89**	m
	150 - 300 mm girth	0.11	1.74	1.44	**3.18**	m
040	Staircases:					
	over 300 mm girth	0.25	3.87	4.54	**8.41**	m2
	not exceeding 150 mm girth	0.08	1.16	0.67	**1.83**	m
	150 - 300 mm girth	0.11	1.74	1.36	**3.10**	m
041	**Prepare; one coat red oxide primer, metalwork surfaces**					
042	General surfaces:					
	over 300 mm girth	0.23	3.52	0.94	**4.46**	m2
	not exceeding 150 mm girth	0.07	1.05	0.14	**1.19**	m
	150 - 300 mm girth	0.11	1.60	0.28	**1.88**	m
	isolated; not exceeding 0.50 m2	0.16	2.44	0.46	**2.90**	Nr
043	Glazed doors and screens in panes:					
	small - not exceeding 0.10 m2	0.58	8.80	0.48	**9.28**	m2
	medium - 0.10 - 0.50 m2	0.41	6.17	0.39	**6.56**	m2
	large - 0.50 - 1.00 m2	0.38	5.82	0.34	**6.16**	m2
	extra large - over 1.00 m2	0.35	5.28	0.30	**5.58**	m2
044	Windows in panes:					
	small - not exceeding 0.10 m2	0.58	8.80	0.39	**9.19**	m2
	medium - 0.10 - 0.50 m2	0.41	6.17	0.34	**6.51**	m2
	large - 0.50 - 1.00 m2	0.38	5.82	0.30	**6.12**	m2
	extra large - over 1.00 m2	0.35	5.28	0.25	**5.53**	m2

Unit Rates

		Man-Hours	Net Labour Price £	Net Mats Price £	Net Unit Price £	Unit
045	Edges of opening casements	0.06	0.88	0.07	**0.95**	m
046	Structural members:					
	over 300 mm girth	0.27	4.05	0.94	**4.99**	m2
	not exceeding 150 mm girth	0.08	1.20	0.14	**1.34**	m
	150 - 300 mm girth	0.12	1.83	0.28	**2.11**	m
047	Each side of ornamental railings, gates and the like (grouped together) measured both sides overall regardless of voids:					
	over 300 mm girth	0.18	2.80	0.65	**3.45**	m2
048	Pipes and conduits, ducting, trunking and the like:					
	over 300 mm girth	0.25	3.87	1.00	**4.87**	m2
	not exceeding 150 mm girth	0.08	1.16	0.15	**1.31**	m
	150 - 300 mm girth	0.11	1.74	0.29	**2.03**	m
049	Eaves gutters:					
	over 300 mm girth	0.25	3.87	1.00	**4.87**	m2
	not exceeding 150 mm girth	0.08	1.16	0.15	**1.31**	m
	150 - 300 mm girth	0.11	1.74	0.29	**2.03**	m
050	Staircases:					
	over 300 mm girth	0.25	3.87	0.94	**4.81**	m2
	not exceeding 150 mm girth	0.08	1.16	0.14	**1.30**	m
	150 - 300 mm girth	0.11	1.74	0.28	**2.02**	m
051	**Prepare; one coat galvanising paint, metalwork surfaces, applied to a dry film thickness of 125 microns**					
052	General surfaces:					
	over 300 mm girth	0.20	3.05	1.90	**4.95**	m2
	not exceeding 150 mm girth	0.06	0.88	0.31	**1.19**	m
	150 - 300 mm girth	0.09	1.37	0.62	**1.99**	m
	isolated; not exceeding 0.50 m2	0.16	2.44	1.37	**3.81**	Nr
053	Glazed doors and screens in panes:					
	small - not exceeding 0.10 m2	0.60	9.11	1.06	**10.17**	m2
	medium - 0.10 - 0.50 m2	0.40	6.06	0.85	**6.91**	m2
	large - 0.50 - 1.00 m2	0.37	5.67	0.74	**6.41**	m2
	extra large - over 1.00 m2	0.33	5.06	0.64	**5.70**	m2
054	Windows in panes:					
	small - not exceeding 0.10 m2	0.60	9.11	0.85	**9.96**	m2
	medium - 0.10 - 0.50 m2	0.40	6.06	0.74	**6.80**	m2
	large - 0.50 - 1.00 m2	0.37	5.67	0.64	**6.31**	m2
	extra large - over 1.00 m2	0.33	5.06	0.54	**5.60**	m2
055	Edges of opening casements	0.07	1.07	0.16	**1.23**	m

Unit Rates

		Man-Hours	Net Labour Price £	Net Mats Price £	Net Unit Price £	Unit
056	Structural members:					
	over 300 mm girth	0.27	4.05	1.90	**5.95**	m2
	not exceeding 150 mm girth	0.08	1.20	0.31	**1.51**	m
	150 - 300 mm girth	0.12	1.83	0.62	**2.45**	m
057	Each side of ornamental railings, gates and the like (grouped together) measured both sides overall regardless of voids:					
	over 300 mm girth	0.16	2.44	1.90	**4.34**	m2
058	Pipes and conduits, ducting, trunking and the like:					
	over 300 mm girth	0.22	3.35	2.23	**5.58**	m2
	not exceeding 150 mm girth	0.07	1.02	0.37	**1.39**	m
	150 - 300 mm girth	0.10	1.49	0.73	**2.22**	m
059	Eaves gutters:					
	over 300 mm girth	0.22	3.35	2.23	**5.58**	m2
	not exceeding 150 mm girth	0.07	1.02	0.37	**1.39**	m
	150 - 300 mm girth	0.10	1.49	0.73	**2.22**	m
060	Staircases:					
	over 300 mm girth	0.27	4.05	1.90	**5.95**	m2
	not exceeding 150 mm girth	0.08	1.20	0.31	**1.51**	m
	150 - 300 mm girth	0.12	1.83	0.62	**2.45**	m
061	**One coat of galvanising paint; second and subsequent coats to previously coated metalwork surfaces; applied to a film thickness of 125 microns**					
062	General surfaces:					
	over 300 mm girth	0.18	2.74	1.58	**4.32**	m2
	not exceeding 150 mm girth	0.05	0.78	0.26	**1.04**	m
	150 - 300 mm girth	0.08	1.19	0.52	**1.71**	m
	isolated; not exceeding 0.50 m2	0.14	2.13	0.78	**2.91**	Nr
063	Glazed doors and screens in panes:					
	small - not exceeding 0.10 m2	0.58	8.83	0.80	**9.63**	m2
	medium - 0.10 - 0.50 m2	0.43	6.49	0.64	**7.13**	m2
	large - 0.50 - 1.00 m2	0.40	6.06	0.57	**6.63**	m2
	extra large - over 1.00 m2	0.36	5.44	0.49	**5.93**	m2
064	Windows in panes:					
	small - not exceeding 0.10 m2	0.58	8.83	0.64	**9.47**	m2
	medium - 0.10 - 0.50 m2	0.43	6.49	0.57	**7.06**	m2
	large - 0.50 - 1.00 m2	0.40	6.06	0.49	**6.55**	m2
	extra large - over 1.00 m2	0.36	5.44	0.41	**5.85**	m2
065	Edges of opening casements	0.06	0.97	0.13	**1.10**	m
066	Structural members:					
	over 300 mm girth	0.25	3.76	1.58	**5.34**	m2
	not exceeding 150 mm girth	0.07	1.10	0.26	**1.36**	m

Unit Rates

		Man-Hours	Net Labour Price £	Net Mats Price £	Net Unit Price £	Unit
	150 - 300 mm girth	0.11	1.66	0.52	**2.18**	m
067	Each side of ornamental railings, gates and the like (grouped together) measured both sides overall regardless of voids:					
	over 300 mm girth	0.14	2.19	1.90	**4.09**	m2
068	Pipes, conduits, ducting, trunking and the like:					
	over 300 mm girth	0.20	3.06	1.86	**4.92**	m2
	not exceeding 150 mm girth	0.06	0.87	0.30	**1.17**	m
	150 - 300 mm girth	0.09	1.43	0.61	**2.04**	m
069	Eaves gutters:					
	over 300 mm girth	0.20	3.06	1.86	**4.92**	m2
	not exceeding 150 mm girth	0.06	0.87	0.30	**1.17**	m
	150 - 300 mm girth	0.08	1.28	0.61	**1.89**	m
070	Staircases:					
	over 300 mm girth	0.23	3.56	1.58	**5.14**	m2
	not exceeding 150 mm girth	0.05	0.70	0.26	**0.96**	m
	150 - 300 mm girth	0.11	1.71	0.52	**2.23**	m
	METALWORK UNDERCOATS - EXTERNALLY					
071	**One undercoat alkyd based paint; primed metalwork surfaces**					
072	General surfaces:					
	over 300 mm girth	0.18	2.80	0.52	**3.32**	m2
	not exceeding 150 mm girth	0.05	0.79	0.07	**0.86**	m
	150 - 300 mm girth	0.08	1.19	0.15	**1.34**	m
	isolated; not exceeding 0.50 m2	0.16	2.44	0.25	**2.69**	Nr
073	Glazed doors and screens in panes:					
	small - not exceeding 0.10 m2	0.52	7.92	0.27	**8.19**	m2
	medium - 0.10 - 0.50 m2	0.35	5.28	0.22	**5.50**	m2
	large - 0.50 - 1.00 m2	0.32	4.92	0.20	**5.12**	m2
	extra large - over 1.00 m2	0.29	4.40	0.17	**4.57**	m2
074	Windows in panes:					
	small - not exceeding 0.10 m2	0.52	7.92	0.22	**8.14**	m2
	medium - 0.10 - 0.50 m2	0.35	5.28	0.20	**5.48**	m2
	large - 0.50 - 1.00 m2	0.32	4.92	0.17	**5.09**	m2
	extra large - over 1.00 m2	0.29	4.40	0.15	**4.55**	m2
075	Edges of opening casements	0.06	0.88	0.04	**0.92**	m
076	Structural members:					
	over 300 mm girth	0.23	3.52	0.52	**4.04**	m2
	not exceeding 150 mm girth	0.07	1.05	0.07	**1.12**	m
	150 - 300 mm girth	0.11	1.60	0.15	**1.75**	m

Unit Rates

		Man-Hours	Net Labour Price £	Net Mats Price £	Net Unit Price £	Unit
077	Each side of ornamental railings, gates and the like (grouped together) measured both sides overall regardless of voids: over 300 mm girth	0.14	2.19	0.42	**2.61**	m2
		0.19	2.89	0.57	**3.46**	m2
		0.06	0.88	0.08	**0.96**	m
		0.09	1.31	0.16	**1.47**	m
		0.19	2.89	0.57	**3.46**	m2
		0.06	0.88	0.08	**0.96**	m
		0.09	1.31	0.15	**1.46**	m
		0.23	3.52	0.52	**4.04**	m2
		0.07	1.05	0.07	**1.12**	m
		0.11	1.60	0.15	**1.75**	m
		0.20	3.00	0.52	**3.52**	m2
		0.06	0.90	0.07	**0.97**	m
		0.09	1.36	0.15	**1.51**	m
		0.16	2.44	0.25	**2.69**	Nr
		0.54	8.27	0.27	**8.54**	m2
		0.37	5.64	0.22	**5.86**	m2
		0.35	5.28	0.20	**5.48**	m2
		0.31	4.75	0.17	**4.92**	m2
		0.54	8.27	0.22	**8.49**	m2
		0.37	5.64	0.20	**5.84**	m2
		0.35	5.28	0.17	**5.45**	m2
		0.31	4.75	0.15	**4.90**	m2
		0.06	0.88	0.04	**0.92**	m
		0.25	3.87	0.52	**4.39**	m2
		0.08	1.16	0.07	**1.23**	m
		0.11	1.74	0.15	**1.89**	m

		Man-Hours	Net Labour Price £	Net Mats Price £	Net Unit Price £	Unit
	Unit Rates					
087	Each side of ornamental railings, gates and the like (grouped together) measured both sides overall regardless of voids:					
	over 300 mm girth	0.16	2.44	0.55	**2.99**	m2
088	Pipes and conduits, ducting, trunking and the like:					
	over 300 mm girth	0.22	3.29	0.57	**3.86**	m2
	not exceeding 150 mm girth	0.07	0.99	0.08	**1.07**	m
	150 - 300 mm girth	0.10	1.48	0.16	**1.64**	m
089	Eaves gutters:					
	over 300 mm girth	0.22	3.29	0.57	**3.86**	m2
	not exceeding 150 mm girth	0.07	0.99	0.08	**1.07**	m
	150 - 300 mm girth	0.10	1.48	0.16	**1.64**	m
090	Staircases:					
	over 300 mm girth	0.24	3.70	0.52	**4.22**	m2
	not exceeding 150 mm girth	0.07	1.11	0.07	**1.18**	m
	150 - 300 mm girth	0.11	1.68	0.15	**1.83**	m
	METALWORK PRIMER FINISH - EXTERNALLY					
091	**Prepare; one coat Hammerite No. 1 primer, metalwork surfaces**					
092	General surfaces:					
	over 300 mm girth	0.19	2.89	0.99	**3.88**	m2
	not exceeding 150 mm girth	0.06	0.87	0.12	**0.99**	m
	150 - 300 mm girth	0.09	1.31	0.25	**1.56**	m
	isolated; not exceeding 0.50 m2	0.06	0.91	0.43	**1.34**	Nr
093	Glazed doors and screens in panes:					
	small - not exceeding 0.10 m2	0.54	8.22	0.45	**8.67**	m2
	medium - 0.10 - 0.50 m2	0.37	5.64	0.35	**5.99**	m2
	large - 0.50 - 1.00 m2	0.35	5.39	0.35	**5.74**	m2
	extra large - over 1.00 m2	0.32	4.83	0.28	**5.11**	m2
094	Windows in panes:					
	small - not exceeding 0.10 m2	0.54	8.22	0.35	**8.57**	m2
	medium - 0.10 - 0.50 m2	0.37	5.64	0.32	**5.96**	m2
	large - 0.50 - 1.00 m2	0.35	5.39	0.28	**5.67**	m2
	extra large - over 1.00 m2	0.32	4.83	0.26	**5.09**	m2
095	Structural members:					
	over 300 mm girth	0.25	3.85	0.75	**4.60**	m2
	not exceeding 150 mm girth	0.08	1.16	0.13	**1.29**	m
	150 - 300 mm girth	0.11	1.74	0.26	**2.00**	m
096	Members of roof trusses:					
	over 300 mm girth	0.32	4.87	1.00	**5.87**	m2
	not exceeding 150 mm girth	0.10	1.55	0.17	**1.72**	m
	150 - 300 mm girth	0.15	2.35	0.33	**2.68**	m

	Unit Rates	Man-Hours	Net Labour Price £	Net Mats Price £	Net Unit Price £	Unit
097	Edges of opening casements	0.06	0.91	0.10	**1.01**	m
098	Each side of ornamental railings, gates and the like (grouped together) measured both sides overall regardless of voids:					
	over 300 mm girth	0.16	2.44	0.90	**3.34**	m2
099	Pipes and conduits, ducting, trunking and the like:					
	over 300 mm girth	0.21	3.18	0.94	**4.12**	m2
	not exceeding 150 mm girth	0.06	0.97	0.17	**1.14**	m
	150 - 300 mm girth	0.09	1.43	0.33	**1.76**	m
100	Staircases:					
	over 300 mm girth	0.25	3.85	0.75	**4.60**	m2
	not exceeding 150 mm girth	0.08	1.16	0.13	**1.29**	m
	150 - 300 mm girth	0.11	1.74	0.26	**2.00**	m
	METALWORK HAMMERED METAL FINISH - EXTERNALLY					
101	**One coat hammered metal finish; primed metalwork surfaces**					
102	General surfaces:					
	over 300 mm girth	0.20	3.05	3.01	**6.06**	m2
	not exceeding 150 mm girth	0.06	0.93	0.45	**1.38**	m
	150 - 300 mm girth	0.09	1.37	1.04	**2.41**	m
	isolated; not exceeding 0.50 m2	0.14	2.13	1.49	**3.62**	Nr
103	Glazed doors and screens in panes:					
	small - not exceeding 0.10 m2	0.58	8.83	1.51	**10.34**	m2
	medium - 0.10 - 0.50 m2	0.37	5.64	1.22	**6.86**	m2
	large - 0.50 - 1.00 m2	0.34	5.21	1.07	**6.28**	m2
	extra large - over 1.00 m2	0.32	4.87	1.22	**6.09**	m2
104	Windows in panes:					
	small - not exceeding 0.10 m2	0.57	8.65	1.22	**9.87**	m2
	medium - 0.10 - 0.50 m2	0.38	5.76	1.07	**6.83**	m2
	large - 0.50 - 1.00 m2	0.35	5.36	0.92	**6.28**	m2
	extra large - over 1.00 m2	0.31	4.75	0.77	**5.52**	m2
105	Structural members:					
	over 300 mm girth	0.25	3.75	3.01	**6.76**	m2
	not exceeding 150 mm girth	0.08	1.20	0.45	**1.65**	m
	150 - 300 mm girth	0.11	1.68	1.04	**2.72**	m
106	Members of roof trusses:					
	over 300 mm girth	0.34	5.18	3.75	**8.93**	m2
	not exceeding 150 mm girth	0.10	1.52	0.60	**2.12**	m
	150 - 300 mm girth	0.15	2.31	1.34	**3.65**	m
107	Edges of opening casements	0.07	1.07	0.12	**1.19**	m

Unit Rates

		Man-Hours	Net Labour Price £	Net Mats Price £	Net Unit Price £	Unit
108	Each side of ornamental railings, gates and the like (grouped together) measured both sides overall regardless of voids:					
	over 300 mm girth	0.18	2.80	2.68	**5.48**	m2
109	Pipes and conduits, ducting, trunking and the like:					
	over 300 mm girth	0.22	3.35	3.46	**6.81**	m2
	not exceeding 150 mm girth	0.06	0.87	0.60	**1.47**	m
	150 - 300 mm girth	0.09	1.34	1.19	**2.53**	m
110	Staircases:					
	over 300 mm girth	0.26	3.90	3.01	**6.91**	m2
	not exceeding 150 mm girth	0.08	1.20	0.45	**1.65**	m
	150 - 300 mm girth	0.11	1.68	1.04	**2.72**	m
111	**Prepare; one coat hammered metal finish, unprimed metalwork surfaces**					
112	General surfaces:					
	over 300 mm girth	0.21	3.20	3.34	**6.54**	m2
	not exceeding 150 mm girth	0.07	1.02	0.55	**1.57**	m
	150 - 300 mm girth	0.10	1.46	1.10	**2.56**	m
	isolated; not exceeding 0.50 m2	0.12	1.83	1.94	**3.77**	Nr
113	Glazed doors and screens in panes:					
	small - not exceeding 0.10 m2	0.56	8.53	1.96	**10.49**	m2
	medium - 0.10 - 0.50 m2	0.39	5.94	1.51	**7.45**	m2
	large - 0.50 - 1.00 m2	0.37	5.70	1.51	**7.21**	m2
	extra large - over 1.00 m2	0.34	5.13	1.22	**6.35**	m2
114	Windows in panes:					
	small - not exceeding 0.10 m2	0.58	8.83	1.51	**10.34**	m2
	medium - 0.10 - 0.50 m2	0.39	5.94	1.37	**7.31**	m2
	large - 0.50 - 1.00 m2	0.37	5.70	1.22	**6.92**	m2
	extra large - over 1.00 m2	0.34	5.13	1.07	**6.20**	m2
115	Structural members:					
	over 300 mm girth	0.27	4.16	3.34	**7.50**	m2
	not exceeding 150 mm girth	0.09	1.31	0.60	**1.91**	m
	150 - 300 mm girth	0.12	1.89	1.19	**3.08**	m
116	Members of roof trusses:					
	over 300 mm girth	0.34	5.18	4.50	**9.68**	m2
	not exceeding 150 mm girth	0.11	1.71	0.75	**2.46**	m
	150 - 300 mm girth	0.16	2.50	1.49	**3.99**	m
117	Edges of opening casements	0.07	1.07	0.12	**1.19**	m
118	Pipes and conduits, ducting, trunking and the like:					
	over 300 mm girth	0.23	3.49	4.20	**7.69**	m2
	not exceeding 150 mm girth	0.08	1.16	0.75	**1.91**	m
	150 - 300 mm girth	0.10	1.58	1.49	**3.07**	m

Unit Rates

		Man-Hours	Net Labour Price £	Net Mats Price £	Net Unit Price £	Unit
119	Staircases:					
	over 300 mm girth	0.27	4.16	3.34	**7.50**	m2
	not exceeding 150 mm girth	0.09	1.31	0.60	**1.91**	m
	150 - 300 mm girth	0.12	1.89	1.19	**3.08**	m
120	Each side of ornamental railings, gates and the like (grouped together) measured both sides overall regardless of voids:					
	over 300 mm girth	0.16	2.44	2.68	**5.12**	m2
	METALWORK INTUMESCENT PAINT - EXTERNALLY					
121	**Prepare; one coat Thermoguard high build primer, primed surfaces**					
122	Structural members:					
	over 300 mm girth	0.27	4.11	1.19	**5.30**	m2
	not exceeding 150 mm girth	0.09	1.31	0.19	**1.50**	m
	150 - 300 mm girth	0.12	1.80	0.36	**2.16**	m
123	Members of roof trusses:					
	over 300 mm girth	0.34	5.24	1.19	**6.43**	m2
	not exceeding 150 mm girth	0.11	1.64	0.19	**1.83**	m
	150 - 300 mm girth	0.16	2.45	0.36	**2.81**	m
124	**One coat 'Thermoguard Thermacoat W' intumescent paint to high build primer, applied by brush, protection given 30 minutes**					
125	Structural members:					
	over 300 mm girth	0.23	3.50	1.71	**5.21**	m2
	not exceeding 150 mm girth	0.07	1.11	0.27	**1.38**	m
	150 - 300 mm girth	0.10	1.52	0.52	**2.04**	m
126	Members of roof trusses:					
	over 300 mm girth	0.29	4.45	1.71	**6.16**	m2
	not exceeding 150 mm girth	0.09	1.40	0.27	**1.67**	m
	150 - 300 mm girth	0.14	2.09	0.52	**2.61**	m
127	**Subsequent coat of 'Thermoguard Thermacoat W' intumescent paint, applied by brush, giving extra protection of 30 minutes per coat up to 2hrs total**					
128	Structural members:					
	over 300 mm girth	0.17	2.63	1.71	**4.34**	m
	not exceeding 150 mm girth	0.06	0.84	0.27	**1.11**	m
	150 - 300 mm girth	0.08	1.14	0.52	**1.66**	m
129	Members of roof trusses:					
	over 300 mm girth	0.22	3.34	1.71	**5.05**	m2
	not exceeding 150 mm girth	0.07	1.05	0.27	**1.32**	m
	150 - 300 mm girth	0.10	1.57	0.52	**2.09**	m

EXTERNAL WORK VL

Unit Rates

		Man-Hours	Net Labour Price £	Net Mats Price £	Net Unit Price £	Unit
VL	**WOODWORK - EXTERNALLY**					
	WOODWORK PRIMERS - EXTERNALLY					
001	**Prepare, knotting, stopping; one coat aluminium wood primer, wood surfaces**					
002	General surfaces:					
	over 300 mm girth	0.23	3.52	0.99	**4.51**	m2
	not exceeding 150 mm girth	0.07	1.05	0.17	**1.22**	m
	150 - 300 mm girth	0.11	1.60	0.29	**1.89**	m
	isolated; not exceeding 0.50 m2	0.16	2.44	0.49	**2.93**	Nr
003	Glazed doors and screens in panes:					
	small - not exceeding 0.10 m2	0.52	7.92	0.61	**8.53**	m2
	medium - 0.10 - 0.50 m2	0.38	5.82	0.51	**6.33**	m2
	large - 0.50 - 1.00 m2	0.35	5.28	0.41	**5.69**	m2
	extra large - over 1.00 m2	0.32	4.92	0.31	**5.23**	m2
004	Windows in panes:					
	small - not exceeding 0.10 m2	0.58	8.80	0.80	**9.60**	m2
	medium - 0.10 - 0.50 m2	0.41	6.17	0.70	**6.87**	m2
	large - 0.50 - 1.00 m2	0.38	5.82	0.62	**6.44**	m2
	extra large - over 1.00 m2	0.35	5.28	0.51	**5.79**	m2
005	Edges of opening casements	0.06	0.88	0.13	**1.01**	m
006	Frames, linings and associated mouldings:					
	over 300 mm girth	0.23	3.52	0.99	**4.51**	m2
	not exceeding 150 mm girth	0.07	1.05	0.17	**1.22**	m
	150 - 300 mm girth	0.11	1.60	0.29	**1.89**	m
007	**Prepare, knotting, stopping; one coat wood primer, wood surfaces**					
008	General surfaces:					
	over 300 mm girth	0.23	3.52	1.03	**4.55**	m2
	not exceeding 150 mm girth	0.07	1.05	0.18	**1.23**	m
	150 - 300 mm girth	0.11	1.60	0.30	**1.90**	m
	isolated; not exceeding 0.50 m2	0.16	2.44	0.51	**2.95**	Nr
009	Glazed doors and screens in panes:					
	small - not exceeding 0.10 m2	0.52	7.92	0.63	**8.55**	m2
	medium - 0.10 - 0.50 m2	0.38	5.82	0.54	**6.36**	m2
	large - 0.50 - 1.00 m2	0.35	5.28	0.42	**5.70**	m2
	extra large - over 1.00 m2	0.32	4.92	0.32	**5.24**	m2
010	Windows in panes:					
	small - not exceeding 0.10 m2	0.58	8.80	0.83	**9.63**	m2
	medium - 0.10 - 0.50 m2	0.41	6.17	0.73	**6.90**	m2
	large - 0.50 - 1.00 m2	0.38	5.82	0.63	**6.45**	m2
	extra large - over 1.00 m2	0.35	5.28	0.53	**5.81**	m2

Unit Rates		Man-Hours	Net Labour Price £	Net Mats Price £	Net Unit Price £	Unit
011	Edges of opening casements	0.06	0.88	0.13	**1.01**	m
012	Frames, linings and associated mouldings:					
	over 300 mm girth	0.23	3.52	1.03	**4.55**	m2
	not exceeding 150 mm girth	0.07	1.05	0.18	**1.23**	m
	150 - 300 mm girth	0.11	1.60	0.30	**1.90**	m
013	**Prepare, knotting, stopping; one coat acrylic wood primer, wood surfaces**					
014	General surfaces:					
	over 300 mm girth	0.23	3.52	1.56	**5.08**	m2
	not exceeding 150 mm girth	0.07	1.05	0.26	**1.31**	m
	150 - 300 mm girth	0.11	1.60	0.43	**2.03**	m
	isolated; not exceeding 0.50 m2	0.16	2.44	0.77	**3.21**	Nr
015	Glazed doors and screens in panes:					
	small - not exceeding 0.10 m2	0.52	7.92	0.95	**8.87**	m2
	medium - 0.10 - 0.50 m2	0.38	5.82	0.79	**6.61**	m2
	large - 0.50 - 1.00 m2	0.35	5.28	0.64	**5.92**	m2
	extra large - over 1.00 m2	0.32	4.92	0.48	**5.40**	m2
016	Windows in panes:					
	small - not exceeding 0.10 m2	0.58	8.80	1.25	**10.05**	m2
	medium - 0.10 - 0.50 m2	0.41	6.17	1.10	**7.27**	m2
	large - 0.50 - 1.00 m2	0.38	5.82	0.95	**6.77**	m2
	extra large - over 1.00 m2	0.35	5.28	0.79	**6.07**	m2
017	Edges of opening casements	0.06	0.88	0.19	**1.07**	m
018	Frames, linings and associated mouldings:					
	over 300 mm girth	0.23	3.52	1.56	**5.08**	m2
	not exceeding 150 mm girth	0.07	1.05	0.26	**1.31**	m
	150 - 300 mm girth	0.11	1.60	0.46	**2.06**	m
	WOODWORK UNDERCOAT - EXTERNALLY					
019	**One undercoat alkyd based paint, white; primed wood surfaces**					
020	General surfaces:					
	over 300 mm girth	0.18	2.80	0.52	**3.32**	m2
	not exceeding 150 mm girth	0.05	0.79	0.07	**0.86**	m
	150 - 300 mm girth	0.08	1.19	0.15	**1.34**	m
	isolated; not exceeding 0.50 m2	0.15	2.28	0.25	**2.53**	Nr
021	Glazed doors and screens in panes:					
	small - not exceeding 0.10 m2	0.46	7.04	0.32	**7.36**	m2
	medium - 0.10 - 0.50 m2	0.32	4.92	0.27	**5.19**	m2
	large - 0.50 - 1.00 m2	0.29	4.40	0.22	**4.62**	m2
	extra large - over 1.00 m2	0.27	4.05	0.17	**4.22**	m2

Unit Rates

		Man-Hours	Net Labour Price £	Net Mats Price £	Net Unit Price £	Unit
022	Windows in panes:					
	small - not exceeding 0.10 m2	0.52	7.92	0.42	**8.34**	m2
	medium - 0.10 - 0.50 m2	0.35	5.28	0.37	**5.65**	m2
	large - 0.50 - 1.00 m2	0.32	4.92	0.32	**5.24**	m2
	extra large - over 1.00 m2	0.29	4.40	0.27	**4.67**	m2
023	Edges of opening casements	0.06	0.88	0.03	**0.91**	m
024	Frames, linings and associated mouldings:					
	over 300 mm girth	0.17	2.65	0.52	**3.17**	m2
	not exceeding 150 mm girth	0.05	0.79	0.07	**0.86**	m
	150 - 300 mm girth	0.08	1.19	0.15	**1.34**	m
025	Signwriting, per 25 mm in height or part thereof, masking and setting out:					
	Arial font	0.16	2.44	0.07	**2.51**	nr
	manifestation or plain logo	0.32	4.87	0.07	**4.94**	nr
	WOODWORK ALKYD BASED GLOSS FINISH - EXTERNALLY					
026	**One coat alkyd based paint, gloss finish; undercoated wood surfaces**					
027	General surfaces:					
	over 300 mm girth	0.20	3.00	0.52	**3.52**	m2
	not exceeding 150 mm girth	0.06	0.90	0.07	**0.97**	m
	150 - 300 mm girth	0.09	1.36	0.15	**1.51**	m
	isolated; not exceeding 0.50 m2	0.15	2.28	0.25	**2.53**	Nr
028	Glazed doors and screens in panes:					
	small - not exceeding 0.10 m2	0.49	7.39	0.32	**7.71**	m2
	medium - 0.10 - 0.50 m2	0.35	5.28	0.27	**5.55**	m2
	large - 0.50 - 1.00 m2	0.31	4.75	0.22	**4.97**	m2
	extra large - over 1.00 m2	0.29	4.40	0.17	**4.57**	m2
029	Windows in panes:					
	small - not exceeding 0.10 m2	0.54	8.27	0.42	**8.69**	m2
	medium - 0.10 - 0.50 m2	0.37	5.64	0.37	**6.01**	m2
	large - 0.50 - 1.00 m2	0.35	5.28	0.32	**5.60**	m2
	extra large - over 1.00 m2	0.31	4.75	0.27	**5.02**	m2
030	Edges of opening casements	0.06	0.88	0.03	**0.91**	m
031	Frames, linings and associated mouldings:					
	over 300 mm girth	0.20	3.00	0.52	**3.52**	m2
	not exceeding 150 mm girth	0.06	0.90	0.07	**0.97**	m
	150 - 300 mm girth	0.09	1.36	0.15	**1.51**	m
032	Signwriting, per 25 mm in height or part thereof, masking and setting out:					
	Arial font	0.10	1.52	0.07	**1.59**	nr
	manifestation or plain logo	0.20	3.05	0.07	**3.12**	nr

EXTERNAL WORK VL

Unit Rates

		Man-Hours	Net Labour Price £	Net Mats Price £	Net Unit Price £	Unit
	VARNISH ON WOODWORK - EXTERNALLY					
033	**Prepare; one coat external grade varnish, wood surfaces, first coat to unprimed wood surfaces**					
034	General surfaces:					
	over 300 mm girth	0.23	3.52	1.39	**4.91**	m2
	not exceeding 150 mm girth	0.07	1.05	0.27	**1.32**	m
	150 - 300 mm girth	0.11	1.60	0.48	**2.08**	m
	isolated; not exceeding 0.50 m2	0.16	2.44	0.73	**3.17**	Nr
035	Glazed doors and screens in panes:					
	small - not exceeding 0.10 m2	0.59	8.99	0.87	**9.86**	m2
	medium - 0.10 - 0.50 m2	0.38	5.82	0.75	**6.57**	m2
	large - 0.50 - 1.00 m2	0.35	5.28	0.59	**5.87**	m2
	extra large - over 1.00 m2	0.32	4.92	0.46	**5.38**	m2
036	Windows in panes:					
	small - not exceeding 0.10 m2	0.64	9.75	1.16	**10.91**	m2
	medium - 0.10 - 0.50 m2	0.41	6.17	1.00	**7.17**	m2
	large - 0.50 - 1.00 m2	0.38	5.82	0.87	**6.69**	m2
	extra large - over 1.00 m2	0.35	5.28	0.75	**6.03**	m2
037	Edges of opening casements	0.06	0.88	0.11	**0.99**	m
038	Frames, linings and associated mouldings:					
	over 300 mm girth	0.23	3.52	1.52	**5.04**	m2
	not exceeding 150 mm girth	0.07	1.05	0.27	**1.32**	m
	150 - 300 mm girth	0.11	1.60	0.53	**2.13**	m
039	**Prepare; one coat external grade varnish, wood surfaces, second and subsequent coats**					
040	General surfaces:					
	over 300 mm girth	0.20	3.00	1.12	**4.12**	m2
	not exceeding 150 mm girth	0.06	0.90	0.16	**1.06**	m
	150 - 300 mm girth	0.09	1.36	0.37	**1.73**	m
	isolated; not exceeding 0.50 m2	0.15	2.28	0.53	**2.81**	Nr
041	Glazed doors and screens in panes:					
	small - not exceeding 0.10 m2	0.56	8.53	0.67	**9.20**	m2
	medium - 0.10 - 0.50 m2	0.35	5.28	0.55	**5.83**	m2
	large - 0.50 - 1.00 m2	0.31	4.75	0.47	**5.22**	m2
	extra large - over 1.00 m2	0.29	4.40	0.35	**4.75**	m2
042	Windows in panes:					
	small - not exceeding 0.10 m2	0.61	9.29	0.88	**10.17**	m2
	medium - 0.10 - 0.50 m2	0.37	5.64	0.79	**6.43**	m2
	large - 0.50 - 1.00 m2	0.35	5.28	0.67	**5.95**	m2
	extra large - over 1.00 m2	0.31	4.75	0.55	**5.30**	m2

Unit Rates

EXTERNAL WORK VL

		Man-Hours	Net Labour Price £	Net Mats Price £	Net Unit Price £	Unit
043	Edges of opening casements	0.06	0.88	0.08	**0.96**	m
044	Frames, linings and associated mouldings:					
	over 300 mm girth	0.20	3.00	1.20	**4.20**	m2
	not exceeding 150 mm girth	0.06	0.90	0.20	**1.10**	m
	150 - 300 mm girth	0.09	1.36	0.37	**1.73**	m
	OIL ON WOODWORK - EXTERNALLY					
045	**Prepare; one coat raw linseed oil, general wood surfaces**					
046	First coat to untreated surfaces:					
	Wrought softwood:					
	over 300 mm girth	0.13	2.01	0.28	**2.29**	m2
	not exceeding 150 mm girth	0.03	0.50	0.04	**0.54**	m
	150 - 300 mm girth	0.06	0.84	0.07	**0.91**	m
	isolated; not exceeding 0.50 m2	0.08	1.22	0.13	**1.35**	Nr
	Sawn softwood:					
	over 300 mm girth	0.17	2.51	0.38	**2.89**	m2
	not exceeding 150 mm girth	0.04	0.67	0.06	**0.73**	m
	150 - 300 mm girth	0.07	1.01	0.09	**1.10**	m
	isolated; not exceeding 0.50 m2	0.10	1.52	0.17	**1.69**	Nr
047	Second and subsequent coats:					
	Wrought softwood:					
	over 300 mm girth	0.13	2.01	0.25	**2.26**	m2
	not exceeding 150 mm girth	0.03	0.50	0.04	**0.54**	m
	150 - 300 mm girth	0.06	0.84	0.07	**0.91**	m
	isolated; not exceeding 0.50 m2	0.08	1.22	0.12	**1.34**	Nr
	Sawn softwood:					
	over 300 mm girth	0.17	2.51	0.33	**2.84**	m2
	not exceeding 150 mm girth	0.04	0.67	0.05	**0.72**	m
	150 - 300 mm girth	0.07	1.01	0.07	**1.08**	m
	isolated; not exceeding 0.50 m2	0.10	1.52	0.15	**1.67**	Nr
048	**Prepare; one coat boiled linseed oil, general wood surfaces**					
049	First coat to untreated surfaces:					
	Wrought softwood:					
	over 300 mm girth	0.13	2.01	0.32	**2.33**	m2
	not exceeding 150 mm girth	0.03	0.50	0.05	**0.55**	m
	150 - 300 mm girth	0.06	0.84	0.08	**0.92**	m
	isolated; not exceeding 0.50 m2	0.08	1.22	0.02	**1.24**	Nr
	Sawn softwood:					
	over 300 mm girth	0.17	2.51	0.43	**2.94**	m2
	not exceeding 150 mm girth	0.04	0.67	0.07	**0.74**	m
	150 - 300 mm girth	0.07	1.01	0.09	**1.10**	m
	isolated; not exceeding 0.50 m2	0.10	1.52	0.20	**1.72**	Nr

Unit Rates

		Man-Hours	Net Labour Price £	Net Mats Price £	Net Unit Price £	Unit
050	Second and subsequent coats:					
	Wrought softwood:					
	over 300 mm girth	0.13	2.01	0.28	**2.29**	m2
	not exceeding 150 mm girth	0.03	0.50	0.05	**0.55**	m
	150 - 300 mm girth	0.06	0.84	0.07	**0.91**	m
	isolated; not exceeding 0.50 m2	0.08	1.22	0.14	**1.36**	Nr
	Sawn softwood:					
	over 300 mm girth	0.17	2.51	0.37	**2.88**	m2
	not exceeding 150 mm girth	0.04	0.67	0.06	**0.73**	m
	150 - 300 mm girth	0.07	1.01	0.08	**1.09**	m
	isolated; not exceeding 0.50 m2	0.10	1.52	0.17	**1.69**	Nr
	Note 051 not used					
	CUPRINOL ON WOODWORK - EXTERNALLY					
052	**Prepare; one coat Cuprinol clear wood preserver, general wood surfaces**					
053	Wrought softwood:					
	over 300 mm girth	0.13	2.01	0.92	**2.93**	m2
	not exceeding 150 mm girth	0.03	0.50	0.11	**0.61**	m
	150 - 300 mm girth	0.06	0.84	0.22	**1.06**	m
	isolated; not exceeding 0.50 m2	0.08	1.22	0.11	**1.33**	Nr
054	Sawn softwood:					
	over 300 mm girth	0.17	2.51	1.49	**4.00**	m2
	not exceeding 150 mm girth	0.04	0.67	0.22	**0.89**	m
	150 - 300 mm girth	0.07	1.01	0.45	**1.46**	m
	isolated; not exceeding 0.50 m2	0.10	1.52	0.22	**1.74**	Nr
055	**Prepare; one coat Cuprinol light oak/dark oak wood preserver, general wood surfaces**					
056	Wrought softwood:					
	over 300 mm girth	0.13	2.01	0.92	**2.93**	m2
	not exceeding 150 mm girth	0.03	0.50	0.11	**0.61**	m
	150 - 300 mm girth	0.06	0.84	0.22	**1.06**	m
	isolated; not exceeding 0.50 m2	0.08	1.22	0.11	**1.33**	Nr
057	Sawn softwood:					
	over 300 mm girth	0.17	2.51	1.49	**4.00**	m2
	not exceeding 150 mm girth	0.04	0.67	0.22	**0.89**	m
	150 - 300 mm girth	0.07	1.01	0.45	**1.46**	m
	isolated; not exceeding 0.50 m2	0.10	1.52	0.22	**1.74**	Nr
058	**Prepare; one coat Cuprinol green wood preserver, general wood surfaces**					
059	Wrought softwood:					
	over 300 mm girth	0.13	2.01	0.92	**2.93**	m2
	not exceeding 150 mm girth	0.03	0.50	0.11	**0.61**	m
	150 - 300 mm girth	0.06	0.84	0.22	**1.06**	m

Unit Rates

		Man-Hours	Net Labour Price £	Net Mats Price £	Net Unit Price £	Unit
	isolated; not exceeding 0.50 m2	0.08	1.22	0.11	**1.33**	Nr
060	Sawn softwood:					
	over 300 mm girth	0.17	2.51	1.49	**4.00**	m2
	not exceeding 150 mm girth	0.04	0.67	0.22	**0.89**	m
	150 - 300 mm girth	0.07	1.01	0.45	**1.46**	m
	isolated; not exceeding 0.50 m2	0.10	1.52	0.22	**1.74**	Nr
061	**Prepare; one coat Cuprinol exterior wood preserver, general wood surfaces**					
062	Wrought softwood:					
	over 300 mm girth	0.13	2.01	0.98	**2.99**	m2
	not exceeding 150 mm girth	0.03	0.50	0.12	**0.62**	m
	150 - 300 mm girth	0.06	0.84	0.24	**1.08**	m
	isolated; not exceeding 0.50 m2	0.08	1.22	0.43	**1.65**	Nr
063	Sawn softwood:					
	over 300 mm girth	0.17	2.51	1.75	**4.26**	m2
	not exceeding 150 mm girth	0.04	0.67	0.21	**0.88**	m
	150 - 300 mm girth	0.07	1.01	0.43	**1.44**	m
	isolated; not exceeding 0.50 m2	0.10	1.52	0.86	**2.38**	Nr
064	**Prepare; one coat Cuprinol red cedar wood preservative, general wood surfaces**					
065	Wrought softwood:					
	over 300 mm girth	0.13	2.01	0.51	**2.52**	m2
	not exceeding 150 mm girth	0.03	0.50	0.08	**0.58**	m
	150 - 300 mm girth	0.06	0.84	0.16	**1.00**	m
	isolated; not exceeding 0.50 m2	0.08	1.22	0.25	**1.47**	Nr
066	Sawn softwood:					
	over 300 mm girth	0.17	2.51	0.92	**3.43**	m2
	not exceeding 150 mm girth	0.04	0.67	0.16	**0.83**	m
	150 - 300 mm girth	0.07	1.01	0.28	**1.29**	m
	isolated; not exceeding 0.50 m2	0.10	1.52	0.41	**1.93**	Nr
067	**Prepare; one coat Cuprinol Premier 5 wood stain, general wood surfaces**					
068	Wrought softwood:					
	over 300 mm girth	0.13	2.01	0.77	**2.78**	m2
	not exceeding 150 mm girth	0.03	0.50	0.10	**0.60**	m
	150 - 300 mm girth	0.06	0.84	0.22	**1.06**	m
	isolated; not exceeding 0.50 m2	0.08	1.22	0.11	**1.33**	Nr
069	Sawn softwood:					
	over 300 mm girth	0.17	2.51	1.04	**3.55**	m2
	not exceeding 150 mm girth	0.04	0.67	0.15	**0.82**	m
	150 - 300 mm girth	0.07	1.01	0.30	**1.31**	m
	isolated; not exceeding 0.50 m2	0.10	1.52	0.15	**1.67**	Nr

Unit Rates

		Man-Hours	Net Labour Price £	Net Mats Price £	Net Unit Price £	Unit
070	**Prepare; one coat Cuprinol Select wood stain, general wood surfaces**					
071	Wrought softwood:					
	over 300 mm girth	0.13	2.01	1.31	**3.32**	m2
	not exceeding 150 mm girth	0.03	0.50	0.43	**0.93**	m
	150 - 300 mm girth	0.06	0.84	0.86	**1.70**	m
	isolated; not exceeding 0.50 m2	0.08	1.22	0.57	**1.79**	Nr
072	Sawn softwood:					
	over 300 mm girth	0.17	2.51	1.75	**4.26**	m2
	not exceeding 150 mm girth	0.04	0.67	0.57	**1.24**	m
	150 - 300 mm girth	0.07	1.01	1.28	**2.29**	m
	isolated; not exceeding 0.50 m2	0.10	1.52	0.57	**2.09**	Nr
	SADOLIN ON WOODWORK - EXTERNALLY					
073	**Prepare; two coats Sadolin Base and two coats Sadolin Extra, general wood surfaces**					
074	To untreated surfaces of wrought softwood:					
	over 300 mm girth	0.53	8.04	1.81	**9.85**	m2
	not exceeding 150 mm girth	0.13	2.01	0.27	**2.28**	m
	150 - 300 mm girth	0.25	3.81	0.51	**4.32**	m
	isolated; not exceeding 0.50 m2	0.32	4.87	0.75	**5.62**	Nr
075	**Prepare; two coats Sadolin Extra, general wood surfaces**					
076	To untreated surfaces of wrought hardwood:					
	over 300 mm girth	0.26	3.96	1.38	**5.34**	m2
	not exceeding 150 mm girth	0.07	0.99	0.22	**1.21**	m
	150 - 300 mm girth	0.11	1.64	0.43	**2.07**	m
	isolated; not exceeding 0.50 m2	0.14	2.13	0.67	**2.80**	Nr
077	**Prepare; one coat Sadolin Base and two coats Sadolin Superdec, general wood surfaces**					
078	To untreated surfaces of wrought softwood:					
	over 300 mm girth	0.39	5.94	2.36	**8.30**	m2
	not exceeding 150 mm girth	0.10	1.48	0.51	**1.99**	m
	150 - 300 mm girth	0.16	2.48	0.88	**3.36**	m
	isolated; not exceeding 0.50 m2	0.20	3.05	1.27	**4.32**	Nr
079	**Prepare; one coat Sadolin Base and two coats Sadolin Classic, general wood surfaces**					
080	To untreated surfaces of wrought softwood:					
	over 300 mm girth	0.39	5.94	2.71	**8.65**	m2
	not exceeding 150 mm girth	0.10	1.48	0.45	**1.93**	m
	150 - 300 mm girth	0.16	2.48	0.73	**3.21**	m
	isolated; not exceeding 0.50 m2	0.20	3.05	1.32	**4.37**	Nr

Unit Rates

		Man-Hours	Net Labour Price £	Net Mats Price £	Net Unit Price £	Unit
VM	**FIBRE CEMENT - EXTERNALLY**					
	FIBRE CEMENT PRIMER - EXTERNALLY					
001	**Prepare; one coat alkali-resisting primer, unprimed fibre cement surfaces**					
002	General surfaces:					
	over 300 mm girth	0.23	3.52	1.81	**5.33**	m2
	not exceeding 150 mm girth	0.07	1.05	0.29	**1.34**	m
	150 - 300 mm girth	0.11	1.60	0.59	**2.19**	m
003	Pipes, conduits, ducting, trunking and the like:					
	over 300 mm girth	0.25	3.87	1.90	**5.77**	m2
	not exceeding 150 mm girth	0.08	1.16	0.31	**1.47**	m
	150 - 300 mm girth	0.11	1.74	0.62	**2.36**	m
004	Eaves gutters:					
	over 300 mm girth	0.25	3.87	1.90	**5.77**	m2
	not exceeding 150 mm girth	0.08	1.16	0.31	**1.47**	m
	150 - 300 mm girth	0.11	1.74	0.62	**2.36**	m
	FIBRE CEMENT UNDERCOATS - EXTERNALLY					
005	**One undercoat alkyd based paint; primed fibre cement surfaces**					
006	General surfaces:					
	over 300 mm girth	0.17	2.63	0.52	**3.15**	m2
	not exceeding 150 mm girth	0.05	0.81	0.07	**0.88**	m
	150 - 300 mm girth	0.08	1.17	0.15	**1.32**	m
007	Pipes, conduits, ducting, trunking and the like:					
	over 300 mm girth	0.19	2.91	0.57	**3.48**	m2
	not exceeding 150 mm girth	0.06	0.88	0.08	**0.96**	m
	150 - 300 mm girth	0.09	1.31	0.16	**1.47**	m
008	Eaves gutters:					
	over 300 mm girth	0.19	2.91	0.57	**3.48**	m2
	not exceeding 150 mm girth	0.06	0.88	0.08	**0.96**	m
	150 - 300 mm girth	0.09	1.31	0.16	**1.47**	m
009	**One coat alkyd based paint, gloss finish; undercoated fibre cement surfaces**					
010	General surfaces:					
	over 300 mm girth	0.20	3.00	0.52	**3.52**	m2
	not exceeding 150 mm girth	0.06	0.88	0.07	**0.95**	m
	150 - 300 mm girth	0.09	1.36	0.15	**1.51**	m
011	Pipes, conduits, ducting, trunking and the like:					
	over 300 mm girth	0.22	3.29	0.57	**3.86**	m2
	not exceeding 150 mm girth	0.07	0.99	0.08	**1.07**	m

Unit Rates

		Man-Hours	Net Labour Price £	Net Mats Price £	Net Unit Price £	Unit
	150 - 300 mm girth	0.10	1.46	0.16	**1.62**	m
012	Eaves gutters:					
	over 300 mm girth	0.22	3.29	0.57	**3.86**	m2
	not exceeding 150 mm girth	0.07	0.99	0.08	**1.07**	m
	150 - 300 mm girth	0.10	1.46	0.16	**1.62**	m
	HARD PAVING - EXTERNALLY					
013	**One coat chlorinated rubber line marking paint thinned as sealer to porous surface**					
014	General surface:					
	over 300 mm girth	0.23	3.50	2.65	**6.15**	m2
	not exceeding 150 mm girth	0.05	0.70	0.40	**1.10**	m
	150 - 300 mm girth	0.09	1.40	0.79	**2.19**	m
	manifestation or numeral 600 mm high	0.65	9.90	1.76	**11.66**	nr
015	**One coat chlorinated rubber line marking paint, including initial setting out**					
016	General surface:					
	over 300 mm girth	0.46	7.01	3.31	**10.32**	m2
	not exceeding 150 mm girth	0.09	1.40	0.50	**1.90**	m
	150 - 300 mm girth	0.18	2.80	0.16	**2.96**	m
	manifestation or numeral 600 mm high	1.30	19.80	2.21	**22.01**	nr
017	**One coat chlorinated rubber line marking paint, each subsequent coat**					
018	General surface:					
	over 300 mm girth	0.23	3.50	3.31	**6.81**	m2
	not exceeding 150 mm girth	0.04	0.53	0.50	**1.03**	m
	150 - 300 mm girth	0.07	1.05	0.16	**1.21**	m
	manifestation or numeral 600 mm high	0.65	9.90	0.08	**9.98**	nr

QSToolbox™
Weapons grade QS software

In these competitive times the systems you employ must keep you on top. There are no prizes for second place in business - coming first is the only option. Lose or win, sink or swim, it's your choice so make it the right one!

QSToolbox™ offers praise winning, profit making solutions for everyone. Drawing measurement - Bill Production - Estimating - Cost Planning, even dimsheet enhancement systems for your Excel™ spreadsheets.

If you want to move up a gear, call us now for your free DVD and brochure pack - it's free. Or go on-line and request your free DVD & brochure pack.

Less effort - less stress - more profits.

Come and see what the future holds.

Go on-line or call now to receive your free demonstration DVD and brochure pack.

Visual Precision
2 Sycamore Tree · Elmhurst Business Park
Park Lane · Elmhurst · Staffs · WS13 8EX
Tel: 01543 262 222 Fax: 01543 262 777
Email: sales@visualprecision.co.uk
Web: http://www.visualprecision.co.uk

CALL NOW | 01543 262 222   **WEB NOW | WWW.VISUALPRECISION.CO.UK**

WALL COVERINGS

RICSBOOKS.COM

Leading supplier of books and contracts for the surveying, construction and property profession.

Shop online at www.ricsbooks.com

- Free delivery on orders over £75
- Exclusive web offers & discounts
- Standard & next day delivery
- Open when you want 24/7

RICS — the mark of property professionalism worldwide | Books

WALL COVERINGS

The prices for "general wallpapers" are given as P.C. prices per standard roll of 5.3 m2. However, due to varying roll lengths and widths of other wall coverings, prices for these are given as P.C. prices per square metre. The P.C. prices are based on the price of an average quality material within each particular range of wall covering.

Waste has been allowed in the constants used for calculating the "Net Materials" prices.

WALL COVERINGS VN

Unit Rates

		Man-Hours	Net Labour Price £	Net Mats Price £	Net Unit Price £	Unit
VN	**DECORATIVE PAPER AND SHEET PLASTIC OR FABRIC LININGS - INTERNALLY**					
001	**Preparing and sizing surfaces prior to hanging paper**					
002	Walls over 300 mm wide:					
	generally	0.09	1.39	0.15	**1.54**	m2
	3.50 - 5.00 m high where ceiling is of dissimilar finish	0.10	1.48	0.15	**1.63**	m2
	staircase areas	0.10	1.48	0.15	**1.63**	m2
003	Ceilings over 300 mm wide:					
	generally	0.11	1.68	0.15	**1.83**	m2
	3.50 - 5.00 m high	0.12	1.78	0.15	**1.93**	m2
	staircase areas	0.12	1.78	0.15	**1.93**	m2
004	**Providing and hanging**					
005	Plain lining paper P.C. £1.90 per roll:					
	walls over 300 mm wide:					
	generally	0.22	3.35	0.59	**3.94**	m2
	3.50 - 5.00 m high where ceiling is of dissimilar finish	0.28	4.19	0.59	**4.78**	m2
	staircase areas	0.28	4.19	0.59	**4.78**	m2
	ceilings over 300 mm wide:					
	generally	0.29	4.34	0.59	**4.93**	m2
	3.50 - 5.00 m high	0.33	5.03	0.59	**5.62**	m2
	staircase areas	0.33	5.03	0.59	**5.62**	m2
006	Woodchip paper P.C. £1.90 per roll:					
	walls over 300 mm wide:					
	generally	0.24	3.69	0.57	**4.26**	m2
	3.50 - 5.00 m high where ceiling is of dissimilar finish	0.30	4.52	0.57	**5.09**	m2
	staircase areas	0.30	4.52	0.57	**5.09**	m2
	ceilings over 300 mm wide:					
	generally	0.31	4.68	0.57	**5.25**	m2
	3.50 - 5.00 m high	0.35	5.36	0.57	**5.93**	m2
	staircase areas	0.35	5.36	0.57	**5.93**	m2
007	Heavy embossed wallpaper P.C. £9.00 per roll:					
	walls over 300 mm wide:					
	generally	0.29	4.42	2.41	**6.83**	m2
	3.50 - 5.00 m high where ceiling is of dissimilar finish	0.35	5.36	2.41	**7.77**	m2
	staircase areas	0.35	5.36	2.41	**7.77**	m2
	ceilings over 300 mm wide:					
	generally	0.36	5.48	2.41	**7.89**	m2
	3.50 - 5.00 m high	0.41	6.20	2.41	**8.61**	m2
	staircase areas	0.41	6.20	2.41	**8.61**	m2

Unit Rates

		Man-Hours	Net Labour Price £	Net Mats Price £	Net Unit Price £	Unit
008	**Providing and hanging decorative paper**					
009	Vinyl surfaced wallpaper P.C. £11.00 per roll:					
	walls over 300 mm wide:					
	generally	0.29	4.48	3.44	**7.92**	m2
	3.50 - 5.00 m high where ceiling is of dissimilar finish	0.35	5.36	3.44	**8.80**	m2
	staircase areas	0.35	5.36	3.44	**8.80**	m2
	ceilings over 300 mm wide:					
	generally	0.36	5.51	3.44	**8.95**	m2
	3.50 - 5.00 m high	0.41	6.20	3.44	**9.64**	m2
	staircase areas	0.41	6.20	3.44	**9.64**	m2
010	Ready pasted vinyl surfaced wallpaper P.C. £14.00 per roll:					
	walls over 300 mm wide:					
	generally	0.27	4.16	4.03	**8.19**	m2
	3.50 - 5.00 m high where ceiling is of dissimilar finish	0.33	5.03	4.03	**9.06**	m2
	staircase areas	0.33	5.03	4.03	**9.06**	m2
	ceilings over 300 mm wide:					
	generally	0.34	5.18	4.03	**9.21**	m2
	3.50 - 5.00 m high	0.39	5.89	4.03	**9.92**	m2
	staircase areas	0.39	5.89	4.03	**9.92**	m2
011	Average quality wallpaper P.C. £12.00 per roll:					
	walls over 300 mm wide:					
	generally	0.29	4.42	3.73	**8.15**	m2
	3.50 - 5.00 m high where ceiling is of dissimilar finish	0.35	5.30	3.73	**9.03**	m2
	staircase areas	0.35	5.30	3.73	**9.03**	m2
	ceilings over 300 mm wide:					
	generally	0.36	5.45	3.73	**9.18**	m2
	3.50 - 5.00 m high	0.41	6.21	3.73	**9.94**	m2
	staircase areas	0.41	6.21	3.73	**9.94**	m2
012	Blown nylon surfaced paper P.C. £7.49 per roll:					
	walls over 300 mm wide:					
	generally	0.30	4.57	2.50	**7.07**	m2
	3.50 - 5.00 m high where ceiling is of dissimilar finish	0.36	5.48	2.50	**7.98**	m2
	staircase areas	0.36	5.48	2.50	**7.98**	m2
	ceilings over 300 mm wide:					
	generally	0.37	5.64	2.50	**8.14**	m2
	3.50 - 5.00 m high	0.42	6.40	2.50	**8.90**	m2
	staircase areas	0.42	6.40	2.50	**8.90**	m2
013	**Providing and hanging wide-width textile wall coverings; to walls with Grade 3 adhesive**					
014	High quality; P.C. £27.50 per square metre:					
	walls over 300 mm wide:					
	generally	0.33	4.95	35.56	**40.51**	m2
	3.50 - 5.00 m high where ceiling is of dissimilar finish	0.38	5.77	35.56	**41.33**	m2
	staircase areas	0.38	5.77	35.56	**41.33**	m2

	Unit Rates	Man-Hours	Net Labour Price £	Net Mats Price £	Net Unit Price £	Unit
015	Mid range; P.C. £20.60 per square metre: walls over 300 mm wide:					
	generally	0.33	4.95	26.79	**31.74**	m2
	3.50 - 5.00 m high where ceiling is of dissimilar finish	0.38	5.77	26.79	**32.56**	m2
	staircase areas	0.38	5.77	26.79	**32.56**	m2
016	**Providing and hanging wide-width glass fibre wall coverings; to walls with Grade 2 adhesive**					
017	P.C. £5.36 per square metre: walls over 300 mm wide:					
	generally	0.30	4.57	7.43	**12.00**	m2
	3.50 - 5.00 m high where ceiling is of dissimilar finish	0.36	5.48	7.43	**12.91**	m2
	staircase areas	0.36	5.48	7.43	**12.91**	m2
018	**Providing and hanging wide-width suede effect wall coverings; to walls with 'Suedefix' adhesive**					
019	P.C. £21.80 per square metre: walls over 300 mm wide:					
	generally	0.38	5.77	28.32	**34.09**	m2
	3.50 - 5.00 m high where ceiling is of dissimilar finish	0.47	7.10	28.32	**35.42**	m2
	staircase areas	0.47	7.10	28.32	**35.42**	m2
020	**Providing and hanging wide-width cotton backed vinyl wall coverings; to walls with Grade 1 adhesive**					
021	P.C. £10.00 per square metre: walls over 300 mm wide:					
	generally	0.28	4.26	13.40	**17.66**	m2
	3.50 - 5.00 m high where ceiling is of dissimilar finish	0.34	5.18	13.40	**18.58**	m2
	staircase areas	0.34	5.18	13.40	**18.58**	m2
022	**Providing and hanging wide-width vinyl wall coverings; to walls with Grade 3 adhesive**					
023	P.C. £2.18 per square metre: walls over 300 mm wide:					
	generally	0.28	4.26	3.19	**7.45**	m2
	3.50 - 5.00 m high where ceiling is of dissimilar finish	0.34	5.18	3.19	**8.37**	m2
	staircase areas	0.34	5.18	3.19	**8.37**	m2

REDECORATION WORK, INTERNALLY

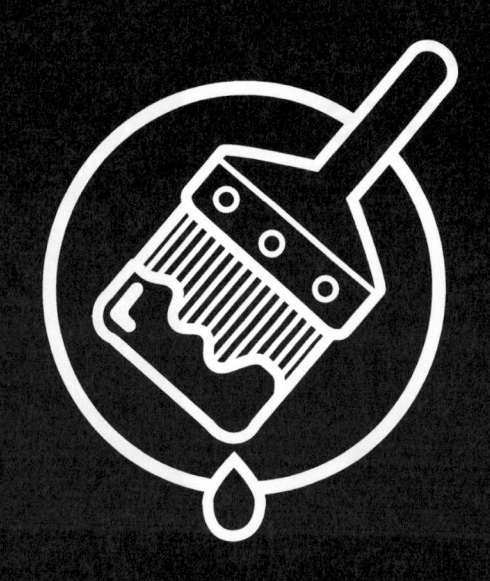

isurv

Building value from knowledge

RICS — the mark of property professionalism worldwide

200 forms, 4000 case studies, unlimited guidance

Yours in a couple of clicks

From Building Surveying to Planning and Valuation. From Commercial Property to Construction and Environment. From Estate Agency to Legal to Sustainability.

isurv gives you the most complete online RICS guidance, ever.

With detailed information on best practice, legal guidance, compliance, case summaries, legislative updates, expert commentary and over 200 downloadable forms; isurv is the definitive source for all property professionals.

And, with more content and more channels, isurv now costs less-per-channel than ever before. Can you afford not to take our free trial?

FREE 7-day trial

Go to **www.isurv.com** or call **024 7686 8433** now

REDECORATION INTERNALLY

Prices in this Section are for preparation of existing surfaces to a condition ready to receive the new paint system. Prices for the new paint system are the same as for painting on new surfaces and they have not been repeated. Refer to New Work Sections and Composite Price Section for new paint prices to obtain total prices for complete redecoration of existing surfaces in accordance with the required specification.

The general descriptions "asbestos cement" and/or "fibre cement" have been used in this book to describe the cementitious board material previously termed "asbestos cement sheeting". Due to the health hazards associated with asbestos fibres, manufacturers of such sheeting materials have largely eliminated the use of asbestos in recent years, hence the term "fibre cement".

However, asbestos products may still be found in use, particularly in older buildings, where redecoration work is required. Particular attention is drawn to the care needed in preparation and treatment of asbestos materials (see "General Information" Section) and if in any doubt, advice should be sought from the Health and Safety Executive whose address and telephone number are given in the "Trade and Government Organisations" Section at the back of this book.

Prices given do NOT include for any additional work or expense which might be incurred in dealing with asbestos surfaces other than that described in Preparation for Painting in the General Information Section of this book.

REDECORATION INTERNAL VO

Unit Rates

		Man-Hours	Net Labour Price £	Net Mats Price £	Net Unit Price £	Unit
VO	**REPAINTING AND REDECORATION WORK**					
	The following prices are for preparation of existing surfaces to a condition ready to receive the new paint system. Prices for the new paint system are the same as for painting on new surfaces and they have not been repeated. See sub-sections VA-VN and VZ for new paint prices to obtain total prices for complete redecoration of existing surfaces in accordance with the required specification.					
	OLD WALLS AND CEILINGS - PREPARATION					
	PREVIOUSLY SIZE DISTEMPERED SURFACES IN ANY CONDITION					
001	**Wash down to remove all traces of old distemper; stop in cracks and rub down; one coat of primer sealer**					
002	Walls over 300 mm wide:					
	plastered	0.40	6.09	1.07	**7.16**	m2
	smooth concrete	0.50	7.63	1.08	**8.71**	m2
	fibre cement	0.44	6.70	1.07	**7.77**	m2
	embossed or textured papered	0.60	9.14	1.07	**10.21**	m2
	cement rendered	0.50	7.63	1.07	**8.70**	m2
	fair face brickwork	0.50	7.63	1.23	**8.86**	m2
	fair face blockwork	0.54	8.22	1.49	**9.71**	m2
003	Walls 3.50 - 5.00 m high where ceiling is of dissimilar finish over 300 mm wide:					
	plastered	0.42	6.40	1.07	**7.47**	m2
	smooth concrete	0.53	8.00	1.08	**9.08**	m2
	fibre cement	0.46	7.04	1.07	**8.11**	m2
	embossed or textured papered	0.62	9.44	1.07	**10.51**	m2
	cement rendered	0.53	8.00	1.07	**9.07**	m2
	fair face brickwork	0.53	8.00	1.23	**9.23**	m2
	fair face blockwork	0.57	8.64	1.49	**10.13**	m2
004	Walls in staircase areas over 300 mm wide:					
	plastered	0.41	6.24	1.07	**7.31**	m2
	smooth concrete	0.51	7.83	1.08	**8.91**	m2
	fibre cement	0.45	6.87	1.07	**7.94**	m2
	embossed or textured papered	0.63	9.59	1.07	**10.66**	m2
	cement rendered	0.51	7.83	1.07	**8.90**	m2
	fair face brickwork	0.51	7.83	1.23	**9.06**	m2
	fair face blockwork	0.55	8.44	1.49	**9.93**	m2
005	Ceilings over 300 mm wide:					
	plastered	0.42	6.40	1.07	**7.47**	m2
	smooth concrete	0.53	8.00	1.08	**9.08**	m2
	fibre cement	0.46	7.04	1.07	**8.11**	m2
	embossed or textured papered	0.64	9.75	1.07	**10.82**	m2
	cement rendered	0.53	8.00	1.07	**9.07**	m2
006	Ceilings 3.50 - 5.00 m high over 300 mm wide:					
	plastered	0.43	6.53	1.07	**7.60**	m2

Unit Rates

		Man-Hours	Net Labour Price £	Net Mats Price £	Net Unit Price £	Unit
	smooth concrete	0.54	8.18	1.08	9.26	m2
	fibre cement	0.47	7.19	1.07	8.26	m2
	embossed or textured papered	0.65	9.90	1.07	10.97	m2
	cement rendered	0.54	8.18	1.07	9.25	m2
007	Ceilings in staircase areas over 300 mm wide:					
	plastered	0.43	6.53	1.07	7.60	m2
	smooth concrete	0.54	8.18	1.08	9.26	m2
	fibre cement	0.47	7.19	1.07	8.26	m2
	embossed or textured papered	0.65	9.90	1.07	10.97	m2
	cement rendered	0.54	8.18	1.07	9.25	m2
	PREVIOUSLY WATER PAINTED, LIME WASHED OR CEMENT PAINTED SURFACES - GOOD CONDITION					
008	**Clean down; stop in cracks and rub down; one coat acrylic primer/undercoat**					
009	Walls over 300 mm wide:					
	plastered	0.24	3.70	1.07	4.77	m2
	smooth concrete	0.31	4.65	1.08	5.73	m2
	fibre cement	0.27	4.05	1.07	5.12	m2
	embossed or textured papered	0.31	4.72	1.07	5.79	m2
	cement rendered	0.31	4.65	1.07	5.72	m2
	fair face brickwork	0.31	4.65	1.23	5.88	m2
	fair face blockwork	0.33	5.00	1.49	6.49	m2
010	Walls 3.50 - 5.00 m high where ceiling is of dissimilar finish; over 300 mm wide:					
	plastered	0.26	3.88	1.07	4.95	m2
	smooth concrete	0.32	4.86	1.08	5.94	m2
	fibre cement	0.28	4.28	1.07	5.35	m2
	embossed or textured papered	0.32	4.87	1.07	5.94	m2
	cement rendered	0.32	4.86	1.07	5.93	m2
	fair face brickwork	0.32	4.86	1.23	6.09	m2
	fair face blockwork	0.34	5.24	1.49	6.73	m2
011	Walls in staircase areas over 300 mm wide:					
	plastered	0.26	3.98	1.07	5.05	m2
	smooth concrete	0.32	4.87	1.08	5.95	m2
	fibre cement	0.28	4.26	1.07	5.33	m2
	embossed or textured papered	0.32	4.87	1.07	5.94	m2
	cement rendered	0.32	4.87	1.07	5.94	m2
	fair face brickwork	0.32	4.87	1.23	6.10	m2
	fair face blockwork	0.34	5.12	1.49	6.61	m2
012	Ceilings over 300 mm wide:					
	plastered	0.26	3.88	1.07	4.95	m2
	smooth concrete	0.32	4.86	1.08	5.94	m2
	fibre cement	0.28	4.28	1.07	5.35	m2
	embossed or textured papered	0.32	4.87	1.07	5.94	m2
	cement rendered	0.32	4.86	1.07	5.93	m2

Unit Rates

		Man-Hours	Net Labour Price £	Net Mats Price £	Net Unit Price £	Unit
013	Ceilings 3.50 - 5.00 m high; over 300 mm wide:					
	plastered	0.27	4.16	1.07	**5.23**	m2
	smooth concrete	0.33	5.03	1.08	**6.11**	m2
	fibre cement	0.29	4.42	1.07	**5.49**	m2
	embossed or textured papered	0.33	5.03	1.07	**6.10**	m2
	cement rendered	0.33	5.03	1.07	**6.10**	m2
014	Ceilings in staircase areas over 300 mm wide:					
	plastered	0.27	4.16	1.07	**5.23**	m2
	smooth concrete	0.33	4.96	1.08	**6.04**	m2
	fibre cement	0.29	4.36	1.07	**5.43**	m2
	embossed or textured papered	0.33	5.03	1.07	**6.10**	m2
	cement rendered	0.33	4.96	1.07	**6.03**	m2

PREVIOUSLY WATER PAINTED, LIME WASHED OR CEMENT PAINTED SURFACES - POOR CONDITION

		Man-Hours	Net Labour Price £	Net Mats Price £	Net Unit Price £	Unit
015	**Remove poorly adhering, flaky, powdery, friable paint; cut back to a firm edge; stop in cracks and rub down; one coat acrylic primer/undercoat**					
016	Walls over 300 mm wide:					
	plastered	0.45	6.85	1.07	**7.92**	m2
	smooth concrete	0.56	8.57	1.08	**9.65**	m2
	fibre cement	0.50	7.54	1.07	**8.61**	m2
	embossed or textured papered	0.66	10.05	1.07	**11.12**	m2
	cement rendered	0.56	8.57	1.07	**9.64**	m2
	fair face brickwork	0.56	8.57	1.23	**9.80**	m2
	fair face blockwork	0.61	9.24	1.49	**10.73**	m2
017	Walls 3.50 - 5.00 m high where ceiling is of dissimilar finish; over 300 mm wide:					
	plastered	0.47	7.20	1.07	**8.27**	m2
	smooth concrete	0.59	9.00	1.08	**10.08**	m2
	fibre cement	0.52	7.92	1.07	**8.99**	m2
	embossed or textured papered	0.68	10.36	1.07	**11.43**	m2
	cement rendered	0.59	9.00	1.07	**10.07**	m2
	fair face brickwork	0.59	9.00	1.23	**10.23**	m2
	fair face blockwork	0.64	9.72	1.49	**11.21**	m2
018	Walls in staircase areas over 300 mm wide:					
	plastered	0.46	7.02	1.07	**8.09**	m2
	smooth concrete	0.58	8.77	1.08	**9.85**	m2
	fibre cement	0.51	7.72	1.07	**8.79**	m2
	embossed or textured papered	0.68	10.36	1.07	**11.43**	m2
	cement rendered	0.58	8.77	1.07	**9.84**	m2
	fair face brickwork	0.58	8.77	1.23	**10.00**	m2
	fair face blockwork	0.62	9.50	1.49	**10.99**	m2
019	Ceilings over 300 mm wide:					
	plastered	0.47	7.20	1.07	**8.27**	m2
	smooth concrete	0.59	9.00	1.08	**10.08**	m2

Unit Rates

REDECORATION INTERNAL VO

		Man-Hours	Net Labour Price £	Net Mats Price £	Net Unit Price £	Unit
	fibre cement	0.52	7.92	1.07	**8.99**	m2
	embossed or textured papered	0.69	10.51	1.07	**11.58**	m2
	cement rendered	0.59	9.00	1.07	**10.07**	m2
020	Ceilings 3.50 - 5.00 m high; over 300 mm wide:					
	plastered	0.48	7.31	1.07	**8.38**	m2
	smooth concrete	0.60	9.14	1.08	**10.22**	m2
	fibre cement	0.53	8.07	1.07	**9.14**	m2
	embossed or textured papered	0.70	10.66	1.07	**11.73**	m2
	cement rendered	0.60	9.14	1.07	**10.21**	m2
021	Ceilings in staircase areas over 300 mm wide:					
	plastered	0.48	7.36	1.07	**8.43**	m2
	smooth concrete	0.60	9.20	1.08	**10.28**	m2
	fibre cement	0.53	8.09	1.07	**9.16**	m2
	embossed or textured papered	0.70	10.66	1.07	**11.73**	m2
	cement rendered	0.60	9.20	1.07	**10.27**	m2
	PREVIOUSLY EMULSION PAINTED SURFACES - GOOD CONDITION					
022	**Clean down; stop in cracks and rub down; one coat acrylic primer/undercoat**					
023	Walls over 300 mm wide:					
	plastered	0.24	3.70	1.07	**4.77**	m2
	smooth concrete	0.31	4.65	1.08	**5.73**	m2
	fibre cement	0.27	4.05	1.07	**5.12**	m2
	embossed or textured papered	0.31	4.72	1.07	**5.79**	m2
	cement rendered	0.31	4.65	1.07	**5.72**	m2
	fair face brickwork	0.31	4.65	1.23	**5.88**	m2
	fair face blockwork	0.33	5.00	1.49	**6.49**	m2
024	Walls 3.50 - 5.00 m high where ceiling is of dissimilar finish; over 300 mm wide:					
	plastered	0.26	3.88	1.07	**4.95**	m2
	smooth concrete	0.32	4.86	1.08	**5.94**	m2
	fibre cement	0.28	4.28	1.07	**5.35**	m2
	embossed or textured papered	0.32	4.87	1.07	**5.94**	m2
	cement rendered	0.32	4.86	1.07	**5.93**	m2
	fair face brickwork	0.32	4.86	1.23	**6.09**	m2
	fair face blockwork	0.34	5.24	1.49	**6.73**	m2
025	Walls in staircase areas over 300 mm wide:					
	plastered	0.26	3.98	1.07	**5.05**	m2
	smooth concrete	0.32	4.87	1.08	**5.95**	m2
	fibre cement	0.28	4.26	1.07	**5.33**	m2
	embossed or textured papered	0.32	4.87	1.07	**5.94**	m2
	cement rendered	0.32	4.87	1.07	**5.94**	m2
	fair face brickwork	0.32	4.87	1.23	**6.10**	m2
	fair face blockwork	0.34	5.12	1.49	**6.61**	m2

Unit Rates

		Man-Hours	Net Labour Price £	Net Mats Price £	Net Unit Price £	Unit
026	Ceilings over 300 mm wide:					
	plastered	0.26	3.88	1.07	**4.95**	m2
	smooth concrete	0.32	4.86	1.08	**5.94**	m2
	fibre cement	0.28	4.28	1.07	**5.35**	m2
	embossed or textured papered	0.32	4.87	1.07	**5.94**	m2
	cement rendered	0.32	4.86	1.07	**5.93**	m2
027	Ceilings 3.50 - 5.00 m high; over 300 mm wide:					
	plastered	0.27	4.11	1.07	**5.18**	m2
	smooth concrete	0.33	4.96	1.08	**6.04**	m2
	fibre cement	0.29	4.36	1.07	**5.43**	m2
	embossed or textured papered	0.33	5.03	1.07	**6.10**	m2
	cement rendered	0.33	4.96	1.07	**6.03**	m2
028	Ceilings in staircase areas over 300 mm wide:					
	plastered	0.27	4.11	1.07	**5.18**	m2
	smooth concrete	0.33	4.96	1.08	**6.04**	m2
	fibre cement	0.29	4.36	1.07	**5.43**	m2
	embossed or textured papered	0.33	5.03	1.07	**6.10**	m2
	cement rendered	0.33	4.96	1.07	**6.03**	m2
	PREVIOUSLY EMULSION PAINTED SURFACES - POOR CONDITION					
029	**Remove poorly adhering, flaky, powdery, friable paint; cut back to a firm edge; stop in cracks and rub down; one coat acrylic primer/undercoat**					
030	Walls over 300 mm wide:					
	plastered	0.37	5.57	1.07	**6.64**	m2
	smooth concrete	0.46	6.99	1.08	**8.07**	m2
	fibre cement	0.41	6.17	1.07	**7.24**	m2
	embossed or textured papered	0.56	8.53	1.07	**9.60**	m2
	cement rendered	0.46	6.99	1.07	**8.06**	m2
	fair face brickwork	0.46	6.99	1.23	**8.22**	m2
	fair face blockwork	0.50	7.54	1.49	**9.03**	m2
031	Walls 3.50 - 5.00 m high where ceiling is of dissimilar finish; over 300 mm wide:					
	plastered	0.38	5.85	1.07	**6.92**	m2
	smooth concrete	0.48	7.34	1.08	**8.42**	m2
	fibre cement	0.43	6.47	1.07	**7.54**	m2
	embossed or textured papered	0.58	8.83	1.07	**9.90**	m2
	cement rendered	0.48	7.34	1.07	**8.41**	m2
	fair face brickwork	0.48	7.34	1.23	**8.57**	m2
	fair face blockwork	0.52	7.92	1.49	**9.41**	m2
032	Walls in staircase areas over 300 mm wide:					
	plastered	0.38	5.71	1.07	**6.78**	m2
	smooth concrete	0.48	7.31	1.08	**8.39**	m2
	fibre cement	0.43	6.55	1.07	**7.62**	m2
	embossed or textured papered	0.59	8.99	1.07	**10.06**	m2
	cement rendered	0.48	7.31	1.07	**8.38**	m2

Unit Rates

		Man-Hours	Net Labour Price £	Net Mats Price £	Net Unit Price £	Unit
	fair face brickwork	0.48	7.31	1.23	**8.54**	m2
	fair face blockwork	0.52	7.92	1.49	**9.41**	m2
033	Ceilings over 300 mm wide:					
	plastered	0.38	5.85	1.07	**6.92**	m2
	smooth concrete	0.48	7.34	1.08	**8.42**	m2
	fibre cement	0.43	6.47	1.07	**7.54**	m2
	embossed or textured papered	0.60	9.14	1.07	**10.21**	m2
	cement rendered	0.48	7.34	1.07	**8.41**	m2
034	Ceilings 3.50 - 5.00 m high; over 300 mm wide:					
	plastered	0.40	6.02	1.07	**7.09**	m2
	smooth concrete	0.49	7.51	1.08	**8.59**	m2
	fibre cement	0.44	6.70	1.07	**7.77**	m2
	embossed or textured papered	0.61	9.29	1.07	**10.36**	m2
	cement rendered	0.49	7.51	1.07	**8.58**	m2
035	Ceilings in staircase areas over 300 mm wide:					
	plastered	0.40	6.02	1.07	**7.09**	m2
	smooth concrete	0.49	7.51	1.08	**8.59**	m2
	fibre cement	0.44	6.70	1.07	**7.77**	m2
	embossed or textured papered	0.61	9.29	1.07	**10.36**	m2
	cement rendered	0.49	7.51	1.07	**8.58**	m2
	PREVIOUSLY ALKYD PAINTED SURFACES - GOOD CONDITION					
036	**Clean down; stop in and rub down; touch in primer and bring forward**					
037	Walls over 300 mm wide:					
	plastered	0.10	1.52	0.04	**1.56**	m2
	smooth concrete	0.13	1.90	0.07	**1.97**	m2
	fibre cement	0.11	1.68	0.04	**1.72**	m2
	embossed or textured papered	0.13	1.90	0.04	**1.94**	m2
	cement rendered	0.13	1.90	0.04	**1.94**	m2
	fair face brickwork	0.13	1.90	0.07	**1.97**	m2
	fair face blockwork	0.14	2.06	0.09	**2.15**	m2
038	Walls 3.50 - 5.00 m high where ceiling is of dissimilar finish; over 300 mm wide:					
	plastered	0.11	1.61	0.04	**1.65**	m2
	smooth concrete	0.14	2.13	0.07	**2.20**	m2
	fibre cement	0.12	1.75	0.04	**1.79**	m2
	embossed or textured papered	0.14	2.13	0.04	**2.17**	m2
	cement rendered	0.14	2.13	0.04	**2.17**	m2
	fair face brickwork	0.14	2.13	0.07	**2.20**	m2
	fair face blockwork	0.15	2.28	0.09	**2.37**	m2
039	Walls in staircase areas over 300 mm wide:					
	plastered	0.11	1.68	0.04	**1.72**	m2
	smooth concrete	0.14	2.13	0.07	**2.20**	m2
	fibre cement	0.12	1.83	0.04	**1.87**	m2

Unit Rates

		Man-Hours	Net Labour Price £	Net Mats Price £	Net Unit Price £	Unit
	embossed or textured papered	0.14	2.13	0.04	**2.17**	m2
	cement rendered	0.14	2.13	0.04	**2.17**	m2
	fair face brickwork	0.14	2.13	0.07	**2.20**	m2
	fair face blockwork	0.15	2.28	0.09	**2.37**	m2
040	Ceilings over 300 mm wide:					
	plastered	0.11	1.68	0.04	**1.72**	m2
	smooth concrete	0.14	2.13	0.07	**2.20**	m2
	fibre cement	0.12	1.83	0.04	**1.87**	m2
	embossed or textured papered	0.14	2.13	0.04	**2.17**	m2
	cement rendered	0.14	2.13	0.04	**2.17**	m2
041	Ceilings 3.50 - 5.00 m high; over 300 mm wide:					
	plastered	0.12	1.83	0.04	**1.87**	m2
	smooth concrete	0.14	2.13	0.07	**2.20**	m2
	fibre cement	0.12	1.80	0.04	**1.84**	m2
	embossed or textured papered	0.14	2.13	0.04	**2.17**	m2
	cement rendered	0.14	2.13	0.04	**2.17**	m2
042	Ceilings in staircase areas over 300 mm wide:					
	plastered	0.11	1.63	0.04	**1.67**	m2
	smooth concrete	0.14	2.13	0.07	**2.20**	m2
	fibre cement	0.12	1.80	0.04	**1.84**	m2
	embossed or textured papered	0.14	2.13	0.04	**2.17**	m2
	cement rendered	0.14	2.13	0.04	**2.17**	m2
	PREVIOUSLY ALKYD PAINTED SURFACES - POOR CONDITION					
043	**Remove old paint completely by burning off; stop in cracks and rub down**					
044	Walls over 300 mm wide:					
	plastered	0.83	12.67	0.11	**12.78**	m2
	smooth concrete	1.04	15.84	0.17	**16.01**	m2
	fibre cement	0.91	13.89	0.11	**14.00**	m2
	cement rendered	1.04	15.84	0.11	**15.95**	m2
	fair face brickwork	1.04	15.84	0.17	**16.01**	m2
	fair face blockwork	1.13	17.13	0.22	**17.35**	m2
045	Walls 3.50 - 5.00 m high where ceiling is of dissimilar finish; over 300 mm wide:					
	plastered	0.88	13.33	0.11	**13.44**	m2
	smooth concrete	1.09	16.63	0.17	**16.80**	m2
	fibre cement	0.96	14.62	0.11	**14.73**	m2
	cement rendered	1.09	16.63	0.11	**16.74**	m2
	fair face brickwork	1.09	16.63	0.17	**16.80**	m2
	fair face blockwork	1.18	18.00	0.22	**18.22**	m2
046	Walls in staircase areas over 300 mm wide:					
	plastered	0.85	12.98	0.11	**13.09**	m2
	smooth concrete	1.07	16.22	0.17	**16.39**	m2
	fibre cement	0.94	14.24	0.11	**14.35**	m2

Unit Rates

		Man-Hours	Net Labour Price £	Net Mats Price £	Net Unit Price £	Unit
	cement rendered	1.07	16.22	0.11	**16.33**	m2
	fair face brickwork	1.07	16.22	0.17	**16.39**	m2
	fair face blockwork	1.15	17.54	0.22	**17.76**	m2
047	Ceilings over 300 mm wide:					
	plastered	0.88	13.33	0.11	**13.44**	m2
	smooth concrete	1.09	16.63	0.17	**16.80**	m2
	fibre cement	0.96	14.62	0.11	**14.73**	m2
	cement rendered	1.09	16.63	0.11	**16.74**	m2
048	Ceilings 3.50 - 5.00 m high; over 300 mm wide:					
	plastered	0.89	13.55	0.11	**13.66**	m2
	smooth concrete	1.12	16.98	0.17	**17.15**	m2
	fibre cement	0.99	15.00	0.11	**15.11**	m2
	cement rendered	1.12	16.98	0.11	**17.09**	m2
049	Ceilings in staircase areas over 300 mm wide:					
	plastered	0.89	13.55	0.11	**13.66**	m2
	smooth concrete	1.12	16.98	0.17	**17.15**	m2
	fibre cement	0.99	15.00	0.11	**15.11**	m2
	cement rendered	1.12	16.98	0.11	**17.09**	m2
050	**Remove old paint completely by chemical means; wash down; stop in cracks and rub down**					
051	Walls over 300 mm wide:					
	plastered	1.00	15.23	3.57	**18.80**	m2
	smooth concrete	1.25	19.04	4.47	**23.51**	m2
	fibre cement	1.10	16.75	3.92	**20.67**	m2
	cement rendered	1.25	19.04	4.45	**23.49**	m2
	fair face brickwork	1.25	19.04	4.47	**23.51**	m2
	fair face blockwork	1.35	20.56	4.83	**25.39**	m2
052	Walls 3.50 - 5.00 m high where ceiling is of dissimilar finish; over 300 mm wide:					
	plastered	1.05	15.98	3.57	**19.55**	m2
	smooth concrete	1.31	20.00	4.47	**24.47**	m2
	fibre cement	1.16	17.59	3.92	**21.51**	m2
	cement rendered	1.31	20.00	4.45	**24.45**	m2
	fair face brickwork	1.31	20.00	4.47	**24.47**	m2
	fair face blockwork	1.42	21.60	4.83	**26.43**	m2
053	Walls in staircase areas over 300 mm wide:					
	plastered	1.02	15.60	3.57	**19.17**	m2
	smooth concrete	1.28	19.52	4.47	**23.99**	m2
	fibre cement	1.13	17.18	3.92	**21.10**	m2
	cement rendered	1.28	19.52	4.45	**23.97**	m2
	fair face brickwork	1.28	19.52	4.47	**23.99**	m2
	fair face blockwork	1.38	21.08	4.83	**25.91**	m2
054	Ceilings over 300 mm wide:					
	plastered	1.05	15.98	3.57	**19.55**	m2

Unit Rates	Man-Hours	Net Labour Price £	Net Mats Price £	Net Unit Price £	Unit
smooth concrete	1.31	20.00	4.47	**24.47**	m2
fibre cement	1.16	17.59	3.92	**21.51**	m2
cement rendered	1.31	20.00	4.45	**24.45**	m2
055 Ceilings 3.50 - 5.00 m high; over 300 mm wide:					
plastered	1.07	16.36	3.57	**19.93**	m2
smooth concrete	1.34	20.42	4.47	**24.89**	m2
fibre cement	1.18	17.97	3.92	**21.89**	m2
cement rendered	1.34	20.42	4.45	**24.87**	m2
056 Ceilings in staircase areas over 300 mm wide:					
plastered	1.07	16.36	3.57	**19.93**	m2
smooth concrete	1.34	20.42	4.47	**24.89**	m2
fibre cement	1.18	17.97	3.92	**21.89**	m2
cement rendered	1.34	20.42	4.45	**24.87**	m2
OLD METALWORK - PREPARATION					
PREVIOUSLY ALKYD PAINTED METALWORK SURFACES - REASONABLY GOOD CONDITION					
057 **Wash down to remove dirt, oil, grease etc.; remove loose and flaking paint; prime bare patches with chromate primer**					
058 General surfaces:					
over 300 mm girth	0.25	3.81	0.12	**3.93**	m2
not exceeding 150 mm girth	0.08	1.22	0.05	**1.27**	m
150 - 300 mm girth	0.11	1.68	0.06	**1.74**	m
isolated; not exceeding 0.50 m2	0.14	2.13	0.09	**2.22**	Nr
059 Glazed doors and screens in panes:					
small - not exceeding 0.10 m2	0.35	5.33	0.10	**5.43**	m2
medium - 0.10 - 0.50 m2	0.30	4.57	0.10	**4.67**	m2
large - 0.50 m2 - 1.00 m2	0.28	4.28	0.06	**4.34**	m2
extra large - over 1.00 m2	0.25	3.81	0.06	**3.87**	m2
060 Windows in panes:					
small - not exceeding 0.10 m2	0.35	5.33	0.10	**5.43**	m2
medium - 0.10 - 0.50 m2	0.30	4.57	0.10	**4.67**	m2
large - 0.50 m2 - 1.00 m2	0.28	4.28	0.06	**4.34**	m2
extra large - over 1.00 m2	0.25	3.81	0.06	**3.87**	m2
061 Structural members:					
over 300 mm girth	0.30	4.57	0.12	**4.69**	m2
not exceeding 150 mm girth	0.09	1.37	0.05	**1.42**	m
150 - 300 mm girth	0.14	2.13	0.06	**2.19**	m
062 Members of roof trusses:					
over 300 mm girth	0.38	5.79	0.12	**5.91**	m2
not exceeding 150 mm girth	0.11	1.68	0.05	**1.73**	m

Unit Rates

		Man-Hours	Net Labour Price £	Net Mats Price £	Net Unit Price £	Unit
	150 - 300 mm girth	0.17	2.60	0.06	**2.66**	m
063	Radiators:					
	over 300 mm girth	0.35	5.33	0.12	**5.45**	m2
064	Pipes and conduits, ducting, trunking and the like:					
	over 300 mm girth	0.28	4.28	0.12	**4.40**	m2
	not exceeding 150 mm girth	0.09	1.37	0.05	**1.42**	m
	150 - 300 mm girth	0.12	1.83	0.06	**1.89**	m
065	Staircases:					
	over 300 mm girth	0.28	4.28	0.12	**4.40**	m2
	not exceeding 150 mm girth	0.09	1.37	0.05	**1.42**	m
	150 - 300 mm girth	0.12	1.83	0.06	**1.89**	m
	Note 066 not used					
	PREVIOUSLY ALKYD PAINTED METALWORK SURFACES - POOR CONDITION					
067	**Remove old paint completely by chemical means; remove rust by wire brushing, chipping and scraping; stop in and rub down**					
068	General surfaces:					
	over 300 mm girth	1.00	15.23	3.59	**18.82**	m2
	not exceeding 150 mm girth	0.30	4.57	1.07	**5.64**	m
	150 - 300 mm girth	0.45	6.85	1.62	**8.47**	m
	isolated; not exceeding 0.50 m2	0.60	9.14	1.79	**10.93**	Nr
069	Glazed doors and screens in panes:					
	small - not exceeding 0.10 m2	1.40	21.32	3.21	**24.53**	m2
	medium - 0.10 - 0.50 m2	1.20	18.28	2.86	**21.14**	m2
	large - 0.50 m2 - 1.00 m2	1.00	15.23	2.50	**17.73**	m2
	extra large - over 1.00 m2	0.80	12.18	2.15	**14.33**	m2
070	Windows in panes:					
	small - not exceeding 0.10 m2	1.40	21.32	3.21	**24.53**	m2
	medium - 0.10 - 0.50 m2	1.20	18.28	2.86	**21.14**	m2
	large - 0.50 m2 - 1.00 m2	1.00	15.23	2.50	**17.73**	m2
	extra large - over 1.00 m2	0.80	12.18	2.15	**14.33**	m2
071	Structural members:					
	over 300 mm girth	1.10	16.75	3.59	**20.34**	m2
	not exceeding 150 mm girth	0.35	5.33	1.07	**6.40**	m
	150 - 300 mm girth	0.50	7.63	1.62	**9.25**	m
072	Members of roof trusses:					
	over 300 mm girth	1.50	22.85	3.59	**26.44**	m2
	not exceeding 150 mm girth	0.45	6.85	1.07	**7.92**	m
	150 - 300 mm girth	0.68	10.36	1.62	**11.98**	m

Unit Rates

		Man-Hours	Net Labour Price £	Net Mats Price £	Net Unit Price £	Unit
073	Radiators:					
	over 300 mm girth	1.40	21.32	3.94	**25.26**	m2
074	Pipes and conduits, ducting, trunking and the like:					
	over 300 mm girth	1.10	16.75	3.59	**20.34**	m2
	not exceeding 150 mm girth	0.35	5.33	1.07	**6.40**	m
	150 - 300 mm girth	0.50	7.63	1.62	**9.25**	m
075	Staircases:					
	over 300 mm girth	1.10	16.75	3.59	**20.34**	m2
	not exceeding 150 mm girth	0.35	5.33	1.07	**6.40**	m
	150 - 300 mm girth	0.50	7.63	1.62	**9.25**	m

Note 076 not used

OLD WOODWORK - PREPARATION

PREVIOUSLY ALKYD PAINTED WOODWORK SURFACES - GOOD CONDITION

		Man-Hours	Net Labour Price £	Net Mats Price £	Net Unit Price £	Unit
077	**Wash down to remove dirt, oil, grease etc.; remove loose and flaking paint; stop in, rub down and prime bare patches**					
078	General surfaces:					
	over 300 mm girth	0.25	3.81	0.04	**3.85**	m2
	not exceeding 150 mm girth	0.08	1.22	0.03	**1.25**	m
	150 - 300 mm girth	0.11	1.68	0.02	**1.70**	m
	isolated; not exceeding 0.50 m2	0.14	2.13	0.02	**2.15**	Nr
079	Glazed doors and screens in panes:					
	small - not exceeding 0.10 m2	0.50	7.63	0.02	**7.65**	m2
	medium - 0.10 - 0.50 m2	0.45	6.85	0.02	**6.87**	m2
	large - 0.50 m2 - 1.00 m2	0.40	6.09	0.02	**6.11**	m2
	extra large - over 1.00 m2	0.35	5.33	0.02	**5.35**	m2
080	Windows in panes:					
	small - not exceeding 0.10 m2	0.55	8.38	0.02	**8.40**	m2
	medium - 0.10 - 0.50 m2	0.50	7.63	0.02	**7.65**	m2
	large - 0.50 m2 - 1.00 m2	0.45	6.85	0.02	**6.87**	m2
	extra large - over 1.00 m2	0.40	6.09	0.02	**6.11**	m2
081	Frames, linings and associated mouldings:					
	over 300 mm girth	0.25	3.81	0.04	**3.85**	m2
	not exceeding 150 mm girth	0.08	1.22	0.01	**1.23**	m
	150 - 300 mm girth	0.11	1.68	0.02	**1.70**	m
082	Cornices:					
	over 300 mm girth	0.30	4.57	0.04	**4.61**	m2
	not exceeding 150 mm girth	0.09	1.37	0.01	**1.38**	m
	150 - 300 mm girth	0.14	2.13	0.02	**2.15**	m

Unit Rates

		Man-Hours	Net Labour Price £	Net Mats Price £	Net Unit Price £	Unit
083	Skirtings, dado rails, picture rails and the like:					
	over 300 mm girth	0.28	4.28	0.04	**4.32**	m2
	not exceeding 150 mm girth	0.09	1.37	0.01	**1.38**	m
	150 - 300 mm girth	0.12	1.83	0.02	**1.85**	m
084	Staircases:					
	over 300 mm girth	0.26	3.96	0.04	**4.00**	m2
	not exceeding 150 mm girth	0.08	1.22	0.01	**1.23**	m
	150 - 300 mm girth	0.12	1.83	0.02	**1.85**	m
	PREVIOUSLY ALKYD PAINTED WOODWORK SURFACES - POOR CONDITION					
085	**Remove old paint completely by burning off; knot, stop in and rub down; one coat wood primer**					
086	General surfaces:					
	over 300 mm girth	0.88	13.33	1.11	**14.44**	m2
	not exceeding 150 mm girth	0.25	3.81	0.20	**4.01**	m
	150 - 300 mm girth	0.38	5.71	0.33	**6.04**	m
	isolated; not exceeding 0.50 m2	0.50	7.62	0.66	**8.28**	Nr
087	Glazed doors and screens in panes:					
	small - not exceeding 0.10 m2	1.62	24.67	0.73	**25.40**	m2
	medium - 0.10 - 0.50 m2	1.50	22.85	0.61	**23.46**	m2
	large - 0.50 m2 - 1.00 m2	1.37	20.87	0.45	**21.32**	m2
	extra large - over 1.00 m2	1.25	19.05	0.32	**19.37**	m2
088	Windows in panes:					
	small - not exceeding 0.10 m2	1.75	26.65	0.91	**27.56**	m2
	medium - 0.10 - 0.50 m2	1.63	24.75	0.81	**25.56**	m2
	large - 0.50 m2 - 1.00 m2	1.50	22.85	0.69	**23.54**	m2
	extra large - over 1.00 m2	1.38	20.94	0.56	**21.50**	m2
089	Frames, linings and associated mouldings:					
	over 300 mm girth	0.88	13.33	1.11	**14.44**	m2
	not exceeding 150 mm girth	0.25	3.81	0.20	**4.01**	m
	150 - 300 mm girth	0.38	5.71	0.33	**6.04**	m
090	Cornices:					
	over 300 mm girth	1.05	15.99	1.11	**17.10**	m2
	not exceeding 150 mm girth	0.32	4.87	0.20	**5.07**	m
	150 - 300 mm girth	0.48	7.23	0.33	**7.56**	m
091	Skirtings, dado rails, picture rails and the like:					
	over 300 mm girth	0.98	14.88	1.11	**15.99**	m2
	not exceeding 150 mm girth	0.30	4.57	0.20	**4.77**	m
	150 - 300 mm girth	0.43	6.50	0.33	**6.83**	m
092	Staircases:					
	over 300 mm girth	0.93	14.09	1.11	**15.20**	m2

Unit Rates

REDECORATION INTERNAL VO

		Man-Hours	Net Labour Price £	Net Mats Price £	Net Unit Price £	Unit
	not exceeding 150 mm girth	0.28	4.19	0.20	**4.39**	m
	150 - 300 mm girth	0.43	6.50	0.33	**6.83**	m
093	**Remove old paint completely by chemical means; wash down; knot, stop in and rub down; one coat wood primer**					
094	General surfaces:					
	over 300 mm girth	1.00	15.23	4.59	**19.82**	m2
	not exceeding 150 mm girth	0.30	4.57	1.25	**5.82**	m
	150 - 300 mm girth	0.45	6.85	1.90	**8.75**	m
	isolated; not exceeding 0.50 m2	0.60	9.14	2.42	**11.56**	Nr
095	Glazed doors and screens in panes:					
	small - not exceeding 0.10 m2	2.00	30.46	3.84	**34.30**	m2
	medium - 0.10 - 0.50 m2	1.80	27.41	3.38	**30.79**	m2
	large - 0.50 m2 - 1.00 m2	1.60	24.35	2.88	**27.23**	m2
	extra large - over 1.00 m2	1.40	21.32	2.43	**23.75**	m2
096	Windows in panes:					
	small - not exceeding 0.10 m2	2.20	33.51	4.36	**37.87**	m2
	medium - 0.10 - 0.50 m2	2.00	30.46	3.94	**34.40**	m2
	large - 0.50 m2 - 1.00 m2	1.80	27.41	3.48	**30.89**	m2
	extra large - over 1.00 m2	1.60	24.35	2.67	**27.02**	m2
097	Frames, linings and associated mouldings:					
	over 300 mm girth	1.00	15.23	4.59	**19.82**	m2
	not exceeding 150 mm girth	0.30	4.57	1.25	**5.82**	m
	150 - 300 mm girth	0.45	6.85	1.90	**8.75**	m
098	Cornices:					
	over 300 mm girth	1.20	18.28	4.59	**22.87**	m2
	not exceeding 150 mm girth	0.36	5.48	1.25	**6.73**	m
	150 - 300 mm girth	0.54	8.22	1.90	**10.12**	m
099	Skirtings, dado rails, picture rails and the like:					
	over 300 mm girth	1.10	16.75	4.59	**21.34**	m2
	not exceeding 150 mm girth	0.33	5.03	1.25	**6.28**	m
	150 - 300 mm girth	0.50	7.63	1.90	**9.53**	m
100	Staircases:					
	over 300 mm girth	1.05	15.98	4.59	**20.57**	m2
	not exceeding 150 mm girth	0.32	4.87	1.25	**6.12**	m
	150 - 300 mm girth	0.47	7.16	1.93	**9.09**	m
	PREVIOUSLY STAINED WOODWORK SURFACES - GOOD CONDITION					
101	**Stop in, rub down and clean off; one coat interior semi-gloss wood stain**					
102	General surfaces:					
	over 300 mm girth	0.15	2.28	0.56	**2.84**	m2

REDECORATION INTERNAL VP

Unit Rates

		Man-Hours	Net Labour Price £	Net Mats Price £	Net Unit Price £	Unit
	not exceeding 150 mm girth	0.05	0.76	0.11	**0.87**	m
	150 - 300 mm girth	0.07	1.07	0.11	**1.18**	m
	isolated; not exceeding 0.50 m2	0.08	1.22	0.21	**1.43**	Nr
103	Glazed doors and screens in panes:					
	small - not exceeding 0.10 m2	0.30	4.57	0.45	**5.02**	m2
	medium - 0.10 - 0.50 m2	0.23	3.50	0.34	**3.84**	m2
	large - 0.50 - 1.00 m2	0.22	3.35	0.24	**3.59**	m2
	extra large - over 1.00 m2	0.21	3.20	0.24	**3.44**	m2
104	Windows in panes:					
	small - not exceeding 0.10 m2	0.33	5.03	0.45	**5.48**	m2
	medium - 0.10 - 0.50 m2	0.25	3.81	0.34	**4.15**	m2
	large - 0.50 - 1.00 m2	0.24	3.66	0.24	**3.90**	m2
	extra large - over 1.00 m2	0.23	3.50	0.24	**3.74**	m2
105	Frames, linings and associated mouldings:					
	over 300 mm girth	0.15	2.28	0.56	**2.84**	m2
	not exceeding 150 mm girth	0.05	0.76	0.11	**0.87**	m
	150 - 300 mm girth	0.07	1.07	0.11	**1.18**	m
106	Skirtings, dado rails, picture rails and the like:					
	over 300 mm girth	0.15	2.28	0.56	**2.84**	m2
	not exceeding 150 mm girth	0.05	0.76	0.11	**0.87**	m
	150 - 300 mm girth	0.07	1.07	0.11	**1.18**	m
107	Staircases:					
	over 300 mm girth	0.15	2.28	0.56	**2.84**	m2
	not exceeding 150 mm girth	0.05	0.76	0.11	**0.87**	m
	150 - 300 mm girth	0.07	1.07	0.11	**1.18**	m

VP — STRIPPING OFF WALLPAPER

		Man-Hours	Net Labour Price £	Net Mats Price £	Net Unit Price £	Unit
001	**Strip off old decorative paper; stop in cracks and rub down**					
002	Lining paper, woodchip paper:					
	one layer from:					
	walls	0.15	2.28	0.07	**2.35**	m2
	ceilings	0.20	3.05	0.07	**3.12**	m2
	two layers from:					
	walls	0.23	3.50	0.07	**3.57**	m2
	ceilings	0.30	4.57	0.07	**4.64**	m2
003	Decorative paper:					
	one layer from:					
	walls	0.18	2.74	0.07	**2.81**	m2
	ceilings	0.24	3.66	0.07	**3.73**	m2
	two layers from:					
	walls	0.27	4.11	0.07	**4.18**	m2
	ceilings	0.36	5.48	0.07	**5.55**	m2

REDECORATION INTERNAL VP

Unit Rates

		Man-Hours	Net Labour Price £	Net Mats Price £	Net Unit Price £	Unit
004	Embossed or textured paper, vinyl surfaced paper:					
	one layer from:					
	walls	0.20	3.05	0.07	**3.12**	m2
	ceilings	0.27	4.11	0.07	**4.18**	m2
	two layers from:					
	walls	0.30	4.57	0.07	**4.64**	m2
	ceilings	0.40	6.09	0.07	**6.16**	m2
005	Anaglypta paper, heavy wallpaper:					
	one layer from:					
	walls	0.30	4.57	0.07	**4.64**	m2
	ceilings	0.40	6.09	0.07	**6.16**	m2
	two layers from:					
	walls	0.40	6.09	0.07	**6.16**	m2
	ceilings	0.55	8.38	0.07	**8.45**	m2

REDECORATION WORK, EXTERNALLY

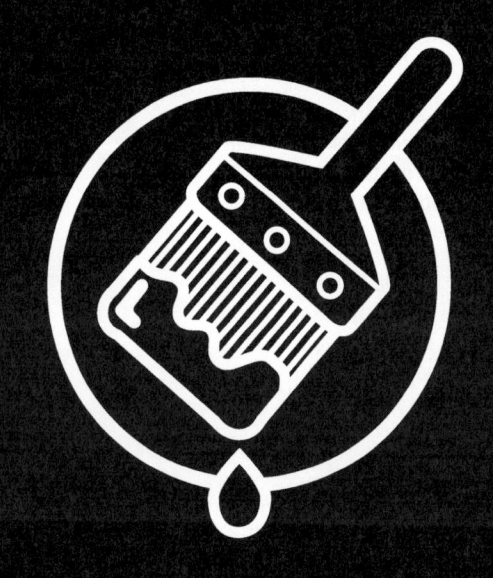

BCIS
50 years celebrating excellence

CONSTRUCTION

LESS DESK TIME MORE FREE TIME

THE REVOLUTION OF THE PRICE BOOK IS HERE
BCIS ONLINE RATES DATABASE

As a purchaser of the 2012 price book, we would like to offer you a free two week trial of our Online Rates Database. It's the price book but online, with lots of additional features to help you to:

- locate prices quickly using faster navigation
- adjust your data to suit a time frame of your choice and choose location factors to make your costs more accurate.

Plus:

- everything you need is in one place – you have a full library at your disposal
- with the full service you can download data into an Excel spreadsheet and manipulate and store your data electronically*.

*During the two week trial data downloads are not available

Offering immediate online access to independent BCIS resource rates data, quantity surveyors and others in the construction industry can have all the information needed to compile and check estimates on their desktops. You won't need to worry about being able to lay your hands on the office copy of the latest price books, all of the information is now easily accessible online.

For a FREE TRIAL of BCIS online rates database, register at **www.bcis.co.uk/ordbdemo**

- Accuracy
- Futureproof
- Value for money
- Saves time
- Flexible
- Customise
- Portable
- Comprehensive

BCIS is the Building Cost Information Service of **RICS** the mark of property professionalism worldwide

REDECORATION EXTERNAL

REDECORATION EXTERNALLY

Prices in this Section are for preparation of existing surfaces to a condition ready to receive the new paint system. Prices for the new paint system are the same as for painting on new surfaces and they have not been repeated. Refer to New Work Sections and Composite Price Section for new paint prices to obtain total prices for complete redecoration of existing surfaces in accordance with the required specification.

The general descriptions "asbestos cement" and/or "fibre cement" have been used in this book to describe the cementitious board material previously termed "asbestos cement sheeting". Due to the health hazards associated with asbestos fibres, manufacturers of such sheeting materials have largely eliminated the use of asbestos in recent years, hence the term "fibre cement".

However, asbestos products may still be found in use, particularly in older buildings, where redecoration work is required. Particular attention is drawn to the care needed in preparation and treatment of asbestos materials (see "General Information" Section) and if in any doubt, advice should be sought from the Health and Safety Executive whose address and telephone number are given in the "Trade and Government Organisations" Section at the back of this book.

Prices given do NOT include for any additional work or expense which might be incurred in dealing with asbestos surfaces other than that described in Preparation for Painting in the General Information Section of this book.

REDECORATION EXTERNAL VQ

Unit Rates

		Man-Hours	Net Labour Price £	Net Mats Price £	Net Unit Price £	Unit
VQ	**REDECORATION WORK EXTERNALLY**					
	PREVIOUSLY EMULSION PAINTED SURFACES - GOOD CONDITION					
001	**Wash down to remove dirt, deposits etc.; stop in cracks, depressions etc. and smooth off; one coat acrylic primer/undercoat**					
002	Walls over 300 mm girth:					
	smooth concrete	0.35	5.33	1.08	**6.41**	m2
	fibre cement	0.30	4.57	1.07	**5.64**	m2
	cement rendered	0.35	5.33	1.07	**6.40**	m2
	rough cast/pebble dash rendered	0.44	6.70	2.92	**9.62**	m2
	Tyrolean rendered	0.70	10.66	3.65	**14.31**	m2
003	Ceilings, beams, soffits etc. over 300 mm girth:					
	smooth concrete	0.38	5.79	1.08	**6.87**	m2
	fibre cement	0.32	4.87	1.07	**5.94**	m2
	cement rendered	0.39	5.94	1.07	**7.01**	m2
	PREVIOUSLY EMULSION PAINTED SURFACES - POOR CONDITION					
004	**Scrape off to remove all flaking or loose material; wash down to remove dirt, deposits etc.; stop in cracks, depressions etc. and smooth off; one coat acrylic primer/undercoat**					
005	Walls over 300 mm girth:					
	smooth concrete	0.51	7.77	1.08	**8.85**	m2
	fibre cement	0.44	6.70	1.07	**7.77**	m2
	cement rendered	0.51	7.77	1.07	**8.84**	m2
	rough cast/pebble dash rendered	0.68	10.36	2.92	**13.28**	m2
	Tyrolean rendered	1.10	16.75	3.65	**20.40**	m2
006	Ceilings, beams, soffits etc. over 300 mm girth:					
	smooth concrete	0.55	8.38	1.08	**9.46**	m2
	fibre cement	0.48	7.31	1.07	**8.38**	m2
	cement rendered	0.56	8.53	1.07	**9.60**	m2
	Note 007 - 009 not used					
	PREVIOUSLY EMULSION PAINTED SURFACES - GOOD CONDITION					
010	**Apply one coat fungicidal solution; wash off (this treatment is in addition to other preparation as required)**					
011	Walls over 300 mm girth:					
	smooth concrete	0.08	1.22	0.21	**1.43**	m2
	fibre cement	0.08	1.22	0.20	**1.42**	m2
	cement rendered	0.09	1.37	0.28	**1.65**	m2
	rough cast/pebble dash rendered	0.10	1.52	0.39	**1.91**	m2
	Tyrolean rendered	0.12	1.83	0.58	**2.41**	m2

REDECORATION EXTERNAL VQ

Unit Rates

		Man-Hours	Net Labour Price £	Net Mats Price £	Net Unit Price £	Unit
012	Ceilings, beams, soffits etc. over 300 mm girth:					
	smooth concrete	0.09	1.37	0.21	**1.58**	m2
	fibre cement	0.09	1.37	0.20	**1.57**	m2
	cement rendered	0.10	1.52	0.28	**1.80**	m2

Note 013 - 019 not used

PREVIOUSLY ALKYD PAINTED METALWORK SURFACES - GOOD CONDITION

		Man-Hours	Net Labour Price £	Net Mats Price £	Net Unit Price £	Unit
020	**Wash down to remove dirt, oil, grease etc.; remove loose and flaking paint; rub down any corroded areas; prime bare patches with zinc phosphate**					
021	General surfaces:					
	over 300 mm girth	0.27	4.11	0.12	**4.23**	m2
	not exceeding 150 mm girth	0.09	1.37	0.02	**1.39**	m
	150 - 300 mm girth	0.12	1.83	0.06	**1.89**	m
	isolated; not exceeding 0.50 m2	0.15	2.28	0.05	**2.33**	Nr
022	Glazed doors and screens in panes:					
	small - not exceeding 0.10 m2	0.37	5.64	0.06	**5.70**	m2
	medium - 0.10 - 0.50 m2	0.32	4.87	0.06	**4.93**	m2
	large - 0.50 - 1.00 m2	0.30	4.57	0.06	**4.63**	m2
	extra large - over 1.00 m2	0.27	4.11	0.06	**4.17**	m2
023	Windows in panes:					
	small - not exceeding 0.10 m2	0.37	5.64	0.06	**5.70**	m2
	medium - 0.10 - 0.50 m2	0.32	4.87	0.06	**4.93**	m2
	large - 0.50 - 1.00 m2	0.30	4.57	0.06	**4.63**	m2
	extra large - over 1.00 m2	0.27	4.11	0.06	**4.17**	m2
024	Eaves gutters:					
	over 300 mm girth	0.30	4.57	0.12	**4.69**	m2
	not exceeding 150 mm girth	0.10	1.52	0.05	**1.57**	m
	150 - 300 mm girth	0.13	1.98	0.06	**2.04**	m
025	Each side of ornamental railings, gates and the like (grouped together) measured both sides overall regardless of voids:					
	over 300 mm girth	0.19	2.92	0.81	**3.73**	m2

Note 026 - 029 not used

PREVIOUSLY ALKYD PAINTED METALWORK SURFACES - POOR CONDITION

		Man-Hours	Net Labour Price £	Net Mats Price £	Net Unit Price £	Unit
030	**Scrape off loose paint and corrosion; wire brush and rub down back to sound or bare metal surface; wash off with white spirit; prime bare patches with zinc phosphate primer**					
031	General surfaces:					
	over 300 mm girth	0.40	6.09	0.28	**6.37**	m2
	not exceeding 150 mm girth	0.13	1.98	0.08	**2.06**	m
	150 - 300 mm girth	0.18	2.74	0.09	**2.83**	m

REDECORATION EXTERNAL VQ

Unit Rates

		Man-Hours	Net Labour Price £	Net Mats Price £	Net Unit Price £	Unit
	isolated; not exceeding 0.50 m2	0.22	3.35	0.13	**3.48**	Nr
032	Glazed doors and screens in panes:					
	small - not exceeding 0.10 m2	0.55	8.38	0.14	**8.52**	m2
	medium - 0.10 - 0.50 m2	0.50	7.62	0.14	**7.76**	m2
	large - 0.50 - 1.00 m2	0.40	6.09	0.09	**6.18**	m2
	extra large - over 1.00 m2	0.35	5.33	0.09	**5.42**	m2
033	Windows in panes:					
	small - not exceeding 0.10 m2	0.55	8.38	0.14	**8.52**	m2
	medium - 0.10 - 0.50 m2	0.50	7.62	0.14	**7.76**	m2
	large - 0.50 - 1.00 m2	0.40	6.09	0.09	**6.18**	m2
	extra large - over 1.00 m2	0.35	5.33	0.09	**5.42**	m2
034	Eaves gutters:					
	over 300 mm girth	1.12	17.06	3.59	**20.65**	m2
	not exceeding 150 mm girth	0.37	5.64	1.07	**6.71**	m
	150 - 300 mm girth	0.53	8.07	1.62	**9.69**	m
035	Each side of ornamental railings, gates and the like (grouped together) measured both sides overall regardless of voids:					
	over 300 mm girth	0.32	4.87	0.81	**5.68**	m2

Note 036 - 039 not used

PREVIOUSLY ALKYD PAINTED WOODWORK SURFACES - GOOD CONDITION

040	**Wash down to remove dirt, oil, grease etc.; scrape off any loose or flaking paint; stop in, rub down and prime bare patches**					
041	General surfaces:					
	over 300 mm girth	0.27	4.11	0.12	**4.23**	m2
	not exceeding 150 mm girth	0.09	1.37	0.02	**1.39**	m
	150 - 300 mm girth	0.12	1.83	0.05	**1.88**	m
	isolated; not exceeding 0.50 m2	0.16	2.44	0.05	**2.49**	Nr
042	Glazed doors and screens in panes:					
	small - not exceeding 0.10 m2	0.54	8.22	0.06	**8.28**	m2
	medium - 0.10 - 0.50 m2	0.48	7.31	0.06	**7.37**	m2
	large - 0.50 - 1.00 m2	0.44	6.70	0.06	**6.76**	m2
	extra large - over 1.00 m2	0.39	5.94	0.06	**6.00**	m2
043	Windows in panes:					
	small - not exceeding 0.10 m2	0.59	8.99	0.06	**9.05**	m2
	medium - 0.10 - 0.50 m2	0.53	8.07	0.06	**8.13**	m2
	large - 0.50 - 1.00 m2	0.49	7.46	0.06	**7.52**	m2
	extra large - over 1.00 m2	0.44	6.70	0.06	**6.76**	m2
044	Frames, linings and associated mouldings:					
	over 300 mm girth	0.27	4.11	0.12	**4.23**	m2

Unit Rates

REDECORATION EXTERNAL VQ

		Man-Hours	Net Labour Price £	Net Mats Price £	Net Unit Price £	Unit
	not exceeding 150 mm girth	0.09	1.37	0.02	**1.39**	m
	150 - 300 mm girth	0.12	1.83	0.05	**1.88**	m
	Note 045 - 049 not used					
	PREVIOUSLY ALKYD PAINTED WOODWORK SURFACES - POOR CONDITION					
050	**Remove old paint completely by burning off; knot, stop in and rub down; one coat wood primer**					
051	General surfaces:					
	over 300 mm girth	0.92	14.01	1.11	**15.12**	m2
	not exceeding 150 mm girth	0.28	4.26	0.20	**4.46**	m
	150 - 300 mm girth	0.42	6.40	0.33	**6.73**	m
	isolated; not exceeding 0.50 m2	0.52	7.92	0.66	**8.58**	Nr
052	Glazed doors and screens in panes:					
	small - not exceeding 0.10 m2	1.66	25.28	0.73	**26.01**	m2
	medium - 0.10 - 0.50 m2	1.54	23.45	0.61	**24.06**	m2
	large - 0.50 - 1.00 m2	1.40	21.32	0.45	**21.77**	m2
	extra large - over 1.00 m2	1.28	19.49	0.32	**19.81**	m2
053	Windows in panes:					
	small - not exceeding 0.10 m2	1.80	27.41	0.91	**28.32**	m2
	medium - 0.10 - 0.50 m2	1.68	25.59	0.81	**26.40**	m2
	large - 0.50 - 1.00 m2	1.54	23.45	0.69	**24.14**	m2
	extra large - over 1.00 m2	1.42	21.63	0.56	**22.19**	m2
054	Frames, linings and associated mouldings					
	over 300 mm girth	0.92	14.01	1.11	**15.12**	m2
	not exceeding 150 mm girth	0.28	4.26	0.20	**4.46**	m
	150 - 300 mm girth	0.34	5.12	0.33	**5.45**	m

RIPAC

The total cost control and contract administration system that maximises performance and flexibility from minimised input

QS's, Consulting Engineers, Contractors, Project Managers, Developers and Client Bodies

- Budget estimates
- Cost planning
- Bills of quantities
- Measurement from CAD / BIM
- Tender pricing and appraisal
- E-tendering
- Resource analysis
- Programme planning links
- Post contract administration
- Payments
- Cash flow
- Whole life costing

Integrated cost control through all project stages.
Easy manipulation of data.
Outputs in user defined formats.
Speedy revisions and updates.
Previous projects available for re-use, analysis and benchmarking.

Various standard libraries of descriptions and price data bases including

BCIS
Independent cost information for the built environment

CSSP CONSTRUCTION SOFTWARE

www.cssp.co.uk

29 London Road
Bromley Kent BR1 1DG
Tel 020 8460 0022
Fax 020 8460 1196
Email enq@cssp.co.uk

COMPOSITE PRICES

BCIS 50 years celebrating excellence

CONSTRUCTION
MAINTENANCE COSTS & REPAIRS
REBUILDING COSTS
INTELLIGENCE
2012

HOW LONG? HOW MUCH?
THE FASTEST, MOST UP-TO-DATE ANSWERS ARE AVAILABLE NOW

Cost information underpins every aspect of the built environment, from construction and rebuilding to maintenance and operation publications.

BCIS, the RICS' Building Cost Information Service, is the leading provider of cost information to the construction industry and anyone else who needs comprehensive, accurate and independent data.

For the past 50 years, BCIS has been collecting, collating, analysing, modelling and interpreting cost information. Today, BCIS make that information easily accessible through online applications, data licensing and publications.

For more information call **+44 (0)870 333 1600** email **contact@bcis.co.uk** or visit **www.bcis.co.uk**

BCIS is the Building Cost Information Service of **RICS**
the mark of property professionalism worldwide

COMPOSITE PRICES

This section offers composite rates for a range of typical specifications for painting work compiled from the earlier sections to facilitate the rapid preparation of estimates.

Unit Rates

COMPOSITE PRICES VZ

VZ COMPOSITE RATES FOR PAINTING. The following is a range of prices compiled from the earlier sections.

WALLS AND CEILINGS - INTERNALLY

EMULSION; VINYL SILK - INTERNALLY

		Man-Hours	Net Labour Price £	Net Mats Price £	Net Unit Price £	Unit
001	One mist and two full coats vinyl silk emulsion paint, white. Emulsion Paint @ £5.44 per 1 ltr					
002	Walls over 300 mm wide:					
	plastered	0.30	4.57	1.43	**6.00**	m2
	smooth concrete	0.38	5.74	1.46	**7.20**	m2
	fibre cement	0.38	5.74	1.75	**7.49**	m2
	embossed or textured papered	0.41	6.20	1.64	**7.84**	m2
	cement rendered	0.34	5.12	1.64	**6.76**	m2
	fair face brickwork	0.46	7.01	1.79	**8.80**	m2
	fair face blockwork	0.53	8.04	2.01	**10.05**	m2
003	Walls 3.50 - 5.00 m high where ceiling is of dissimilar finish; over 300 mm wide:					
	plastered	0.35	5.39	1.43	**6.82**	m2
	smooth concrete	0.40	6.02	1.46	**7.48**	m2
	fibre cement	0.40	6.02	1.75	**7.77**	m2
	embossed or textured papered	0.43	6.47	1.64	**8.11**	m2
	cement rendered	0.35	5.39	1.64	**7.03**	m2
	fair face brickwork	0.48	7.33	1.79	**9.12**	m2
	fair face blockwork	0.56	8.45	2.01	**10.46**	m2
004	Walls in staircase areas over 300 mm wide:					
	plastered	0.34	5.22	1.43	**6.65**	m2
	smooth concrete	0.39	5.91	1.46	**7.37**	m2
	fibre cement	0.39	5.91	1.75	**7.66**	m2
	embossed or textured papered	0.42	6.37	1.64	**8.01**	m2
	cement rendered	0.34	5.22	1.64	**6.86**	m2
	fair face brickwork	0.47	7.16	1.79	**8.95**	m2
	fair face blockwork	0.54	8.25	2.01	**10.26**	m2
005	Ceilings over 300 mm wide:					
	plastered	0.35	5.39	1.43	**6.82**	m2
	smooth concrete	0.40	6.02	1.46	**7.48**	m2
	fibre cement	0.40	6.02	1.75	**7.77**	m2
	embossed or textured papered	0.43	6.47	1.64	**8.11**	m2
	cement rendered	0.35	5.39	1.64	**7.03**	m2
006	Ceilings 3.50 - 5.00 m high over 300 mm wide:					
	plastered	0.37	5.67	1.43	**7.10**	m2
	smooth concrete	0.42	6.34	1.46	**7.80**	m2
	fibre cement	0.42	6.34	1.75	**8.09**	m2
	embossed or textured papered	0.45	6.79	1.64	**8.43**	m2
	cement rendered	0.37	5.67	1.64	**7.31**	m2

	Unit Rates	Man-Hours	Net Labour Price £	Net Mats Price £	Net Unit Price £	Unit
007	Ceilings in staircase areas over 300 mm wide:					
	plastered	0.36	5.50	1.43	**6.93**	m2
	smooth concrete	0.41	6.18	1.46	**7.64**	m2
	fibre cement	0.41	6.18	1.75	**7.93**	m2
	embossed or textured papered	0.44	6.64	1.64	**8.28**	m2
	cement rendered	0.36	5.50	1.64	**7.14**	m2
	PRIMERS - INTERNALLY					
008	**ADD to the foregoing if with alkali-resisting primer in lieu of mist coat.** Primer @ £12.50 per 1 ltr					
009	Walls and Ceilings:					
	plastered	0.07	1.01	1.13	**2.14**	m2
	smooth concrete	0.07	1.05	1.13	**2.18**	m2
	fibre cement	0.07	1.05	1.13	**2.18**	m2
	cement rendered	0.07	1.01	1.51	**2.52**	m2
	fair face brickwork	0.08	1.25	1.51	**2.76**	m2
	fair face blockwork	0.10	1.52	2.15	**3.67**	m2
	ALKYD BASED PAINT; EGGSHELL FINISH - INTERNALLY					
010	**One coat of all-purpose primer; and two coats eggshell finish alkyd paint.** Primer @ £15.07 per 1 ltr; Eggshell @ £6.51 per 1 ltr					
011	Walls over 300 mm wide:					
	plastered	0.39	5.88	1.37	**7.25**	m2
	smooth concrete	0.43	6.53	1.41	**7.94**	m2
	fibre cement	0.43	6.61	2.06	**8.67**	m2
	embossed or textured papered	0.46	6.99	1.81	**8.80**	m2
	cement rendered	0.39	5.88	2.03	**7.91**	m2
	fair face brickwork	0.48	7.31	2.17	**9.48**	m2
	fair face blockwork	0.56	8.59	2.49	**11.08**	m2
012	Walls 3.50 - 5.00 m high where ceiling is of dissimilar finish; over 300 mm wide:					
	plastered	0.41	6.21	1.37	**7.58**	m2
	smooth concrete	0.46	6.94	1.96	**8.90**	m2
	fibre cement	0.46	6.94	2.06	**9.00**	m2
	embossed or textured papered	0.48	7.31	1.81	**9.12**	m2
	cement rendered	0.41	6.21	2.03	**8.24**	m2
	fair face brickwork	0.50	7.62	2.17	**9.79**	m2
	fair face blockwork	0.58	8.83	2.49	**11.32**	m2
013	Walls in staircase areas over 300 mm wide:					
	plastered	0.40	6.05	1.37	**7.42**	m2
	smooth concrete	0.45	6.79	1.96	**8.75**	m2
	fibre cement	0.45	6.79	2.06	**8.85**	m2
	embossed or textured papered	0.48	7.25	1.81	**9.06**	m2
	cement rendered	0.40	6.05	2.03	**8.08**	m2
	fair face brickwork	0.50	7.62	2.17	**9.79**	m2

Unit Rates

COMPOSITE PRICES VZ

		Man-Hours	Net Labour Price £	Net Mats Price £	Net Unit Price £	Unit
	fair face blockwork	0.58	8.83	2.49	**11.32**	m2
014	Ceilings over 300 mm wide:					
	plastered	0.41	6.21	1.37	**7.58**	m2
	smooth concrete	0.46	6.94	1.96	**8.90**	m2
	fibre cement	0.46	6.94	2.06	**9.00**	m2
	embossed or textured papered	0.48	7.31	1.81	**9.12**	m2
	cement rendered	0.41	6.21	2.03	**8.24**	m2
015	Ceilings 3.50 - 5.00 m high over 300 mm wide:					
	plastered	0.43	6.53	1.37	**7.90**	m2
	smooth concrete	0.48	7.28	1.96	**9.24**	m2
	fibre cement	0.48	7.28	2.06	**9.34**	m2
	embossed or textured papered	0.50	7.66	1.81	**9.47**	m2
	cement rendered	0.43	6.53	2.03	**8.56**	m2
016	Ceilings in staircase areas over 300 mm wide:					
	plastered	0.42	6.35	1.37	**7.72**	m2
	smooth concrete	0.47	7.13	1.96	**9.09**	m2
	fibre cement	0.47	7.13	2.06	**9.19**	m2
	embossed or textured papered	0.49	7.52	1.75	**9.27**	m2
	cement rendered	0.42	6.35	2.03	**8.38**	m2
	ALKYD BASED PAINT; GLOSS FINISH - INTERNALLY					
017	**One coat of all-purpose primer; one coat alkyd based undercoat; one coat alkyd based gloss finish.** Primer @ £15.07 ; Undercoat @ £5.97 ; Gloss @ £5.97 (all per 1 ltr)					
018	Walls over 300 mm wide:					
	plastered	0.39	5.94	1.29	**7.23**	m2
	smooth concrete	0.44	6.63	1.63	**8.26**	m2
	fibre cement	0.44	6.63	1.68	**8.31**	m2
	embossed or textured papered	0.46	7.04	1.91	**8.95**	m2
	cement rendered	0.39	5.94	1.91	**7.85**	m2
	fair face brickwork	0.50	7.58	2.15	**9.73**	m2
	fair face blockwork	0.58	8.82	2.47	**11.29**	m2
019	Walls 3.50 - 5.00 m high where ceiling is of dissimilar finish; over 300 mm wide:					
	plastered	0.41	6.24	1.29	**7.53**	m2
	smooth concrete	0.46	6.96	1.63	**8.59**	m2
	fibre cement	0.46	6.96	1.68	**8.64**	m2
	embossed or textured papered	0.48	7.28	1.91	**9.19**	m2
	cement rendered	0.41	6.24	1.91	**8.15**	m2
	fair face brickwork	0.52	7.95	2.15	**10.10**	m2
	fair face blockwork	0.61	9.26	2.47	**11.73**	m2
020	Walls in staircase areas over 300 mm wide:					
	plastered	0.40	6.08	1.29	**7.37**	m2
	smooth concrete	0.45	6.81	1.63	**8.44**	m2
	fibre cement	0.45	6.81	1.68	**8.49**	m2

Unit Rates	Man-Hours	Net Labour Price £	Net Mats Price £	Net Unit Price £	Unit
embossed or textured papered	0.47	7.17	1.91	**9.08**	m2
cement rendered	0.40	6.08	1.91	**7.99**	m2
fair face brickwork	0.51	7.78	2.15	**9.93**	m2
fair face blockwork	0.60	9.14	2.47	**11.61**	m2
021 Ceilings over 300 mm wide:					
plastered	0.43	6.49	1.29	**7.78**	m2
smooth concrete	0.46	6.96	1.63	**8.59**	m2
fibre cement	0.46	6.96	1.68	**8.64**	m2
embossed or textured papered	0.50	7.62	1.91	**9.53**	m2
cement rendered	0.41	6.24	1.91	**8.15**	m2
022 Ceilings 3.50 - 5.00 m high over 300 mm wide:					
plastered	0.44	6.70	1.29	**7.99**	m2
smooth concrete	0.47	7.20	1.63	**8.83**	m2
fibre cement	0.47	7.20	1.68	**8.88**	m2
embossed or textured papered	0.51	7.77	1.91	**9.68**	m2
cement rendered	0.43	6.49	1.91	**8.40**	m2
023 Ceilings in staircase areas over 300 mm wide:					
plastered	0.44	6.70	1.29	**7.99**	m2
smooth concrete	0.47	7.14	1.63	**8.77**	m2
fibre cement	0.47	7.14	1.68	**8.82**	m2
embossed or textured papered	0.51	7.77	1.91	**9.68**	m2
cement rendered	0.43	6.55	1.91	**8.46**	m2
ANTI-BACTERIAL PAINT SYSTEM - INTERNALLY					
024 **Prepare; apply one coat primer/adhesive, two coats water based hygiene paint, high performance two component epoxy coating to**					
025 Walls over 300 mm wide:					
plastered	0.32	4.79	6.99	**11.78**	m2
smooth concrete	0.35	5.29	6.99	**12.28**	m2
fibre cement	0.38	5.81	7.43	**13.24**	m2
embossed or textured paper	0.38	5.81	7.43	**13.24**	m2
cement render	0.42	6.41	7.43	**13.84**	m2
fair face brickwork	0.46	7.01	7.43	**14.44**	m2
fair face blockwork	0.51	7.73	7.75	**15.48**	m2
026 Walls 3.50 - 5.00 m high where ceiling is of dissimilar finish; over 300 mm wide:					
plastered	0.34	5.17	7.26	**12.43**	m2
smooth concrete	0.37	5.69	7.11	**12.80**	m2
fibre cement	0.41	6.26	7.43	**13.69**	m2
embossed or textured paper	0.42	6.32	7.43	**13.75**	m2
cement render	0.45	6.89	7.43	**14.32**	m2
fair face brickwork	0.50	7.55	7.43	**14.98**	m2
fair face blockwork	0.55	8.31	7.90	**16.21**	m2

Unit Rates

		Man-Hours	Net Labour Price £	Net Mats Price £	Net Unit Price £	Unit
027	Walls in staircase area over 300 mm wide:					
	plastered	0.37	5.57	6.99	**12.56**	m2
	smooth concrete	0.40	6.10	6.99	**13.09**	m2
	fibre cement	0.44	6.73	7.43	**14.16**	m2
	embossed or textured paper	0.31	4.72	7.43	**12.15**	m2
	cement render	0.49	7.41	7.43	**14.84**	m2
	fair face brickwork	0.53	8.13	7.43	**15.56**	m2
	fair face blockwork	0.59	8.93	7.90	**16.83**	m2
028	Ceilings over 300 mm wide:					
	plastered	0.35	5.29	6.99	**12.28**	m2
	smooth concrete	0.38	5.81	6.99	**12.80**	m2
	fibre cement	0.42	6.41	7.43	**13.84**	m2
	cement render	0.46	7.01	7.43	**14.44**	m2
029	Ceilings 3.50 - 5.00 m high over 300 mm wide:					
	plastered	0.35	5.39	6.99	**12.38**	m2
	smooth concrete	0.39	5.92	6.99	**12.91**	m2
	fibre cement	0.43	6.53	7.43	**13.96**	m2
	cement render	0.47	7.14	7.43	**14.57**	m2
030	Ceilings in staircase areas over 300 mm wide:					
	plastered	0.39	5.93	6.99	**12.92**	m2
	smooth concrete	0.43	6.50	6.99	**13.49**	m2
	fibre cement	0.47	7.17	7.38	**14.55**	m2
	cement render	0.52	7.84	7.38	**15.22**	m2
	INTUMESCENT PAINT SYSTEMS - INTERNALLY					
031	**Prepare; one coat Thermoguard Wallcoat and one coat Thermoguard flame retardant acrylic matt, designated class O surface spread of frame**					
032	Walls over 300 mm wide:					
	plastered	0.24	3.66	3.79	**7.45**	m2
	smooth concrete	0.26	4.02	4.07	**8.09**	m2
	fibre cement	0.29	4.42	4.24	**8.66**	m2
	embossed or textured paper	0.24	3.68	4.24	**7.92**	m2
	cement render	0.32	4.85	4.24	**9.09**	m2
	fair face brickwork	0.35	5.34	4.24	**9.58**	m2
	fair face blockwork	0.39	5.88	4.80	**10.68**	m2
033	Walls 3.50 - 5.00 m high where ceiling is of dissimilar finish; over 300 mm wide:					
	plastered	0.25	3.86	3.79	**7.65**	m2
	smooth concrete	0.28	4.33	4.07	**8.40**	m2
	fibre cement	0.31	4.75	4.24	**8.99**	m2
	embossed or textured paper	0.26	3.96	4.24	**8.20**	m2
	cement render	0.34	5.22	4.24	**9.46**	m2
	fair face brickwork	0.38	5.74	4.24	**9.98**	m2
	fair face blockwork	0.42	6.32	4.45	**10.77**	m2

Unit Rates

		Man-Hours	Net Labour Price £	Net Mats Price £	Net Unit Price £	Unit
034	Walls in staircase area over 300 mm wide:					
	plastered	0.27	4.14	3.79	**7.93**	m2
	smooth concrete	0.31	4.66	4.07	**8.73**	m2
	fibre cement	0.34	5.12	4.24	**9.36**	m2
	embossed or textured paper	0.28	4.26	4.24	**8.50**	m2
	cement render	0.37	5.62	4.24	**9.86**	m2
	fair face brickwork	0.41	6.17	4.24	**10.41**	m2
	fair face blockwork	0.45	6.80	4.45	**11.25**	m2
035	Ceilings over 300 mm wide:					
	plastered	0.26	4.02	4.07	**8.09**	m2
	smooth concrete	0.29	4.42	4.07	**8.49**	m2
	fibre cement	0.32	4.85	4.24	**9.09**	m2
	cement render	0.35	5.34	4.24	**9.58**	m2
036	Ceilings 3.50 - 5.00 m high over 300 mm wide:					
	plastered	0.28	4.21	3.79	**8.00**	m2
	smooth concrete	0.31	4.64	4.07	**8.71**	m2
	fibre cement	0.34	5.10	4.24	**9.34**	m2
	cement render	0.37	5.62	4.24	**9.86**	m2
037	Ceilings in staircase areas over 300 mm wide:					
	plastered	0.29	4.44	4.07	**8.51**	m2
	smooth concrete	0.32	4.87	4.07	**8.94**	m2
	fibre cement	0.35	5.36	4.24	**9.60**	m2
	cement render	0.39	5.90	4.24	**10.14**	m2
038	**Prepare; one coat Thermoguard Wallcoat and two coats Thermoguard flame retardant acrylic eggshell, designated class O surface spread of frame**					
039	Walls over 300 mm wide:					
	plastered	0.35	5.34	5.36	**10.70**	m2
	smooth concrete	0.39	5.86	5.64	**11.50**	m2
	fibre cement	0.42	6.45	5.98	**12.43**	m2
	embossed or textured paper	0.36	5.52	5.98	**11.50**	m2
	cement render	0.47	7.07	5.98	**13.05**	m2
	fair face brickwork	0.51	7.79	5.98	**13.77**	m2
	fair face blockwork	0.56	8.58	6.75	**15.33**	m2
040	Walls 3.50 - 5.00 m high where ceiling is of dissimilar finish; over 300 mm wide:					
	plastered	0.37	5.63	5.36	**10.99**	m2
	smooth concrete	0.41	6.31	5.64	**11.95**	m2
	fibre cement	0.46	6.93	5.98	**12.91**	m2
	embossed or textured paper	0.39	5.94	5.98	**11.92**	m2
	cement render	0.50	7.61	5.98	**13.59**	m2
	fair face brickwork	0.55	8.37	5.98	**14.35**	m2
	fair face blockwork	0.61	9.21	6.40	**15.61**	m2
041	Walls in staircase area over 300 mm wide:					
	plastered	0.40	6.04	5.36	**11.40**	m2

Unit Rates

		Man-Hours	Net Labour Price £	Net Mats Price £	Net Unit Price £	Unit
	smooth concrete	0.45	6.79	5.64	**12.43**	m2
	fibre cement	0.49	7.47	5.98	**13.45**	m2
	embossed or textured paper	0.42	6.39	5.98	**12.37**	m2
	cement render	0.54	8.19	5.98	**14.17**	m2
	fair face brickwork	0.59	9.00	5.98	**14.98**	m2
	fair face blockwork	0.65	9.91	6.40	**16.31**	m2
042	Ceilings over 300 mm wide:					
	plastered	0.39	5.86	5.64	**11.50**	m2
	smooth concrete	0.42	6.45	5.64	**12.09**	m2
	fibre cement	0.47	7.07	5.98	**13.05**	m2
	cement render	0.51	7.79	5.98	**13.77**	m2
	fair face brickwork	0.56	8.58	5.98	**14.56**	m2
	fair face blockwork	0.62	9.44	6.40	**15.84**	m2
043	Ceilings 3.50 - 5.00 m high over 300 mm wide:					
	plastered	0.40	6.14	5.36	**11.50**	m2
	smooth concrete	0.45	6.77	5.64	**12.41**	m2
	fibre cement	0.49	7.43	5.98	**13.41**	m2
	cement render	0.54	8.19	5.98	**14.17**	m2
	fair face brickwork	0.59	9.00	5.98	**14.98**	m2
	fair face blockwork	0.65	9.93	6.40	**16.33**	m2
044	Ceilings in staircase areas over 300 mm wide:					
	plastered	0.42	6.47	5.64	**12.11**	m2
	smooth concrete	0.47	7.11	5.64	**12.75**	m2
	fibre cement	0.51	7.81	5.98	**13.79**	m2
	cement render	0.56	8.60	5.98	**14.58**	m2
	fair face brickwork	0.62	9.44	5.98	**15.42**	m2
	fair face blockwork	0.68	10.41	6.40	**16.81**	m2
	METALWORK - INTERNALLY					
045	**Prepare; one coat zinc phosphate primer, one undercoat alkyd based paint, one coat alkyd based paint gloss finish**					
046	General surfaces:					
	over 300 mm girth	0.58	8.80	1.98	**10.78**	m2
	not exceeding 150 mm girth	0.17	2.63	0.28	**2.91**	m
	150 - 300 mm girth	0.26	3.98	0.57	**4.55**	m
	isolated; not exceeding 0.50 m2	0.30	4.57	0.95	**5.52**	Nr
047	Glazed doors and screens in panes:					
	small - not exceeding 0.10 m2	1.58	24.03	1.02	**25.05**	m2
	medium - 0.10 - 0.50 m2	1.08	16.42	0.84	**17.26**	m2
	large - 0.50 - 1.00 m2	1.01	15.40	0.74	**16.14**	m2
	extra large - over 1.00 m2	0.91	13.87	0.64	**14.51**	m2
048	Windows in panes:					
	small - not exceeding 0.10 m2	1.58	24.03	0.84	**24.87**	m2

Unit Rates

		Man-Hours	Net Labour Price £	Net Mats Price £	Net Unit Price £	Unit
	medium - 0.10 - 0.50 m2	1.08	16.42	0.74	**17.16**	m2
	large - 0.50 - 1.00 m2	1.01	15.40	0.64	**16.04**	m2
	extra large - over 1.00 m2	0.91	13.87	0.54	**14.41**	m2
049	Structural members:					
	over 300 mm girth	0.72	11.00	1.98	**12.98**	m2
	not exceeding 150 mm girth	0.22	3.30	0.28	**3.58**	m
	150 - 300 mm girth	0.33	4.96	0.57	**5.53**	m
050	Members of roof trusses:					
	over 300 mm girth	0.94	14.38	2.60	**16.98**	m2
	not exceeding 150 mm girth	0.28	4.31	0.38	**4.69**	m
	150 - 300 mm girth	0.43	6.49	0.77	**7.26**	m
051	Radiators:					
	over 300 mm girth	0.83	12.69	2.20	**14.89**	m2
052	Pipes and conduits, ducting, trunking and the like:					
	over 300 mm girth	0.64	9.69	2.12	**11.81**	m2
	not exceeding 150 mm girth	0.19	2.91	0.34	**3.25**	m
	150 - 300 mm girth	0.29	4.36	0.63	**4.99**	m
053	Staircases:					
	over 300 mm girth	0.70	10.66	1.98	**12.64**	m2
	not exceeding 150 mm girth	0.21	3.20	0.28	**3.48**	m
	150 - 300 mm girth	0.32	4.81	0.57	**5.38**	m
054	**Prepare; one coat Thermoguard high build primer, one coat 'Thermoguard Thermacoat W' intumescent paint to high build primer, protection given 30 minutes, second coat Thermoguard flame retardant acrylic matt/eggshell**					
055	Structural members:					
	over 300 mm girth	0.70	10.67	4.38	**15.05**	m2
	not exceeding 150 mm girth	0.25	3.81	0.68	**4.49**	m
	150 - 300 mm girth	0.35	5.26	1.30	**6.56**	m
056	Members of roof trusses:					
	over 300 mm girth	0.77	11.73	4.42	**16.15**	m2
	not exceeding 150 mm girth	0.27	4.07	0.68	**4.75**	m
	150 - 300 mm girth	0.37	5.62	1.33	**6.95**	m
057	**Prepare; one coat Thermoguard high build primer, two coats 'Thermoguard Thermacoat W' intumescent paint to high build primer, protection given 60 minutes, third coat Thermoguard flame retardant acrylic matt/eggshell**					
058	Structural members:					
	over 300 mm girth	0.88	13.41	6.12	**19.53**	m2
	not exceeding 150 mm girth	0.33	5.03	0.95	**5.98**	m
	150 - 300 mm girth	0.46	6.94	1.82	**8.76**	m

Unit Rates

		Man-Hours	Net Labour Price £	Net Mats Price £	Net Unit Price £	Unit
059	Members of roof trusses:					
	over 300 mm girth	0.97	14.75	6.13	**20.88**	m2
	not exceeding 150 mm girth	0.36	5.41	0.95	**6.36**	m
	150 - 300 mm girth	0.49	7.46	1.85	**9.31**	m
060	**One coat wood primer, one coat Thermo Guard Timber, one coat flame retardant matt paint, giving class O spread of flame, to timber surface**					
061	General surfaces:					
	over 300 mm girth	0.52	7.92	5.44	**13.36**	m2
	not exceeding 150 mm girth	0.14	2.17	0.84	**3.01**	m
	150 - 300 mm girth	0.22	3.38	1.62	**5.00**	m
	isolated; not exceeding 0.5 m2	0.32	4.85	2.41	**7.26**	m
062	Fire resistant glazed doors, screens and windows:					
	small - not exceeding 0.10 m2	1.31	20.02	13.73	**33.75**	m2
	medium - 0.10 - 0.50 m2	0.98	14.93	10.40	**25.33**	m2
	large - 0.50 - 1 m2	0.85	12.98	8.95	**21.93**	m2
	extra large - over 1.00 m2	0.71	10.82	7.54	**18.36**	m2
	opening edge	0.14	2.15	0.88	**3.03**	m
063	Staircases:					
	over 300 mm girth	0.54	8.26	7.81	**16.07**	m2
	not exceeding 150 mm girth	0.15	2.30	1.16	**3.46**	m
	150 - 300 mm girth	0.24	3.59	2.33	**5.92**	m
	isolated; not exceeding 0.5 m2	0.30	4.56	3.41	**7.97**	nr
	balustrade (measured both sides)	0.36	5.47	29.27	**34.74**	m2
	WOODWORK - INTERNALLY					
064	**Prepare, knotting, stopping; one coat wood primer, one undercoat alkyd based paint, one coat alkyd based paint gloss finish**					
065	General surfaces:					
	over 300 mm girth	0.58	8.80	2.08	**10.88**	m2
	not exceeding 150 mm girth	0.17	2.63	0.29	**2.92**	m
	150 - 300 mm girth	0.26	3.98	0.58	**4.56**	m
	isolated; not exceeding 0.50 m2	0.30	4.57	1.00	**5.57**	Nr
066	Glazed doors and screens in panes:					
	small - not exceeding 0.10 m2	1.51	23.01	1.29	**24.30**	m2
	medium - 0.10 - 0.50 m2	1.01	15.38	1.06	**16.44**	m2
	large - 0.50 - 1.00 m2	0.91	13.87	0.86	**14.73**	m2
	extra large - over 1.00 m2	0.84	12.85	0.64	**13.49**	m2
067	Windows in panes:					
	small - not exceeding 0.10 m2	1.58	24.03	1.64	**25.67**	m2
	medium - 0.10 - 0.50 m2	1.08	16.42	1.51	**17.93**	m2
	large - 0.50 - 1.00 m2	0.98	14.89	1.29	**16.18**	m2
	extra large - over 1.00 m2	0.91	13.87	1.06	**14.93**	m2

Unit Rates

		Man-Hours	Net Labour Price £	Net Mats Price £	Net Unit Price £	Unit
068	Frames, linings and associated mouldings:					
	over 300 mm girth	0.58	8.80	2.08	**10.88**	m2
	not exceeding 150 mm girth	0.17	2.63	0.29	**2.92**	m
	150 - 300 mm girth	0.26	3.98	0.58	**4.56**	m
069	Cornices:					
	over 300 mm girth	0.69	10.49	2.08	**12.57**	m2
	not exceeding 150 mm girth	0.21	3.14	0.29	**3.43**	m
	150 - 300 mm girth	0.32	4.81	0.58	**5.39**	m
070	Skirtings, dado rails, picture rails and the like:					
	over 300 mm girth	0.64	9.69	2.08	**11.77**	m2
	not exceeding 150 mm girth	0.19	2.91	0.29	**3.20**	m
	150 - 300 mm girth	0.29	4.40	0.58	**4.98**	m
071	Staircases:					
	over 300 mm girth	0.61	9.26	2.08	**11.34**	m2
	not exceeding 150 mm girth	0.18	2.77	0.29	**3.06**	m
	150 - 300 mm girth	0.27	4.17	0.58	**4.75**	m
072	**One coat wood primer, one coat Thermo Guard Timber, two coats flame retardant eggshell paint, giving class O spread of flame, to timber surface**					
073	General surfaces:					
	over 300 mm girth	0.63	9.60	7.01	**16.61**	m2
	not exceeding 150 mm girth	0.16	2.42	1.07	**3.49**	m
	150 - 300 mm girth	0.26	3.89	2.09	**5.98**	m
	CLEAR FINISHES ON WOODWORK - INTERNALLY					
074	**Prepare; two coats polyurethane varnish, first coat on unprimed surfaces.** Varnish @£11.68 ; Thinners @ £3.75 (Both per 1 ltr)					
075	General surfaces:					
	over 300 mm girth	0.41	6.24	1.83	**8.07**	m2
	not exceeding 150 mm girth	0.12	1.87	0.32	**2.19**	m
	150 - 300 mm girth	0.19	2.83	0.61	**3.44**	m
	isolated; not exceeding 0.50 m2	0.24	3.66	0.91	**4.57**	Nr
076	Glazed doors and screens in panes:					
	small - not exceeding 0.10 m2	0.97	14.73	1.13	**15.86**	m2
	medium - 0.10 - 0.50 m2	0.70	10.66	0.95	**11.61**	m2
	large - 0.50 - 1.00 m2	0.63	9.64	0.77	**10.41**	m2
	extra large - over 1.00 m2	0.59	8.97	0.60	**9.57**	m2
077	Windows in panes:					
	small - not exceeding 0.10 m2	1.08	16.42	1.48	**17.90**	m2
	medium - 0.10 - 0.50 m2	0.74	11.33	1.30	**12.63**	m2
	large - 0.50 - 1.00 m2	0.70	10.66	1.13	**11.79**	m2
	extra large - over 1.00 m2	0.63	9.64	0.95	**10.59**	m2

Unit Rates

		Man-Hours	Net Labour Price £	Net Mats Price £	Net Unit Price £	Unit
078	Frames, linings and associated mouldings:					
	over 300 mm girth	0.41	6.26	1.97	**8.23**	m2
	not exceeding 150 mm girth	0.12	1.87	0.35	**2.22**	m
	150 - 300 mm girth	0.19	2.83	0.64	**3.47**	m
079	Cornices:					
	over 300 mm girth	0.49	7.45	1.97	**9.42**	m2
	not exceeding 150 mm girth	0.15	2.22	0.35	**2.57**	m
	150 - 300 mm girth	0.22	3.41	0.64	**4.05**	m
080	Skirtings, dado rails, picture rails and the like:					
	over 300 mm girth	0.45	6.88	1.97	**8.85**	m2
	not exceeding 150 mm girth	0.14	2.07	0.35	**2.42**	m
	150 - 300 mm girth	0.21	3.14	0.64	**3.78**	m
081	Staircases:					
	over 300 mm girth	0.43	6.58	1.83	**8.41**	m2
	not exceeding 150 mm girth	0.13	1.98	0.32	**2.30**	m
	150 - 300 mm girth	0.20	2.99	0.64	**3.63**	m
082	**Prepare; two coats Thermoguard fire varnish on timber, giving Class O flame spread**					
083	General surfaces:					
	over 300 mm girth	0.41	6.24	4.14	**10.38**	m2
	not exceeding 150 mm girth	0.13	1.96	0.62	**2.58**	m
	150 - 300 mm girth	0.19	2.95	1.24	**4.19**	m
	isolated; not exceeding 0.5 m2	0.25	3.75	1.87	**5.62**	nr
084	Fire resistant glazed doors, screens and windows:					
	small - not exceeding 0.10 m2	1.05	15.99	2.79	**18.78**	m2
	medium - 0.10 - 0.50 m2	0.76	11.58	2.30	**13.88**	m2
	large - 0.50 - 1 m2	0.66	10.06	1.88	**11.94**	m2
	extra large - over 1 m2	0.59	8.91	10.93	**19.84**	m2
	opening edge	0.14	2.14	0.44	**2.58**	m
085	Staircases:					
	over 300 mm girth	0.43	6.55	2.76	**9.31**	m2
	not exceeding 150 mm girth	0.14	2.06	0.62	**2.68**	m
	150 - 300 mm girth	0.21	3.20	1.24	**4.44**	m
	isolated; not exceeding 0.5 m2	0.27	4.11	1.86	**5.97**	m
	balustrade (measured both sides)	0.36	5.48	3.53	**9.01**	m2
086	**Prepare; one thin coat sealer, one full coat quick drying floor, one coat anti-slip paint to**					
087	Floor, concrete or the like:					
	over 300 mm girth	0.30	4.49	1.31	**5.80**	m2
	not exceeding 150 mm girth	0.08	1.13	0.26	**1.39**	m
	150 - 300 mm girth	0.11	1.59	0.40	**1.99**	m

Unit Rates

		Man-Hours	Net Labour Price £	Net Mats Price £	Net Unit Price £	Unit
	NEW WORK EXTERNALLY					
	WALLS - EXTERNALLY					
	EMULSION PAINT - WALLS EXTERNALLY					
088	**Two coats of exterior emulsion paint, matt finish, first coat to unprimed surfaces.** Exterior emulsion @ £4.69 per 1 ltr					
089	To walls:					
	smooth concrete	0.28	4.31	0.86	**5.17**	m2
	fibre cement	0.28	4.31	0.84	**5.15**	m2
	cement rendered	0.26	3.93	0.98	**4.91**	m2
	fair face brickwork	0.34	5.24	1.10	**6.34**	m2
	fair face blockwork	0.40	6.05	1.26	**7.31**	m2
	rough cast/pebbledash rendered	0.53	8.01	2.79	**10.80**	m2
	Tyrolean rendered	0.68	10.37	3.49	**13.86**	m2
	MASONRY PAINT - WALLS EXTERNALLY					
090	**One coat of masonry sealer, to unprimed surfaces and two coats masonry paint (white).** Masonry sealer @ £9.03 ; Masonry paint @ £9.20 (Both per 1 ltr)					
091	To walls:					
	smooth concrete	0.40	6.09	2.56	**8.65**	m2
	fibre cement	0.40	6.09	2.53	**8.62**	m2
	cement rendered	0.36	5.53	2.89	**8.42**	m2
	fair face brickwork	0.49	7.46	3.02	**10.48**	m2
	fair face blockwork	0.56	8.57	2.77	**11.34**	m2
	rough cast/pebbledash rendered	0.76	11.56	5.20	**16.76**	m2
	Tyrolean rendered	0.97	14.82	7.83	**22.65**	m2
	SNOWCEM PAINT - WALLS EXTERNALLY					
092	**Two coats of 'Snowcem' cement paint; including base coat of stabilising solution to unprimed surface.** Stabilising solution @ £6.64 per 1 ltr; 'Snowcem' @ £1.34 per 1 kg					
093	To walls:					
	smooth concrete	0.40	6.09	0.76	**6.85**	m2
	fibre cement	0.40	6.09	0.73	**6.82**	m2
	cement rendered	0.36	5.51	0.81	**6.32**	m2
	fair face brickwork	0.49	7.46	0.85	**8.31**	m2
	fair face blockwork	0.56	8.53	1.05	**9.58**	m2
	rough cast/pebbledash rendered	0.76	11.56	1.42	**12.98**	m2
	Tyrolean rendered	0.97	14.80	2.87	**17.67**	m2

Unit Rates

		Man-Hours	Net Labour Price £	Net Mats Price £	Net Unit Price £	Unit
	SANDTEX MATT - WALLS EXTERNALLY					
094	**Two coats 'Sandtex' (Fine build on Matt), including base coat of stabilising solution to unprimed surface.** Stabilising solution @ £6.64 per 1 ltr; 'Sandtex' Matt @ £3.73 per 1 ltr; 'Sandtex' Fine Build @ £2.93 per 1 kg					
095	To walls:					
	smooth concrete	0.51	7.71	5.25	**12.96**	m2
	fibre cement	0.51	7.71	5.21	**12.92**	m2
	cement rendered	0.47	7.13	5.36	**12.49**	m2
	fair face brickwork	0.62	9.47	5.39	**14.86**	m2
	fair face blockwork	0.70	10.62	5.60	**16.22**	m2
	WALLS, SPRAYED FINISHES - EXTERNALLY					
	EMULSION PAINT, MATT SPRAYED - EXTERNALLY					
096	**Two coats of emulsion paint, matt finish, first coat to unprimed surface.** Exterior emulsion @ £4.69 per 1 ltr					
097	Walls over 300 mm wide:					
	smooth concrete	0.10	2.38	1.08	**3.46**	m2
	fibre cement	0.10	2.38	1.22	**3.60**	m2
	cement rendered	0.11	2.65	1.22	**3.87**	m2
	fair face brickwork	0.13	3.03	1.55	**4.58**	m2
	fair face blockwork	0.14	3.24	1.55	**4.79**	m2
	rough cast/pebbledash rendered	0.15	3.48	1.97	**5.45**	m2
	Tyrolean rendered	0.17	3.98	7.04	**11.02**	m2
	METALWORK - EXTERNALLY					
098	**Prepare; one coat zinc phosphate primer, one undercoat alkyd based paint, one coat alkyd based paint gloss finish**					
099	General surfaces:					
	over 300 mm girth	0.61	9.31	1.98	**11.29**	m2
	not exceeding 150 mm girth	0.18	2.73	0.28	**3.01**	m
	150 - 300 mm girth	0.27	4.07	0.57	**4.64**	m
	isolated; not exceeding 0.50 m2	0.36	5.48	0.95	**6.43**	Nr
100	Glazed doors and screens in panes:					
	small - not exceeding 0.10 m2	1.64	25.04	1.02	**26.06**	m2
	medium - 0.10 - 0.50 m2	1.14	17.42	0.84	**18.26**	m2
	large - 0.50 - 1.00 m2	1.07	16.25	0.74	**16.99**	m2
	extra large - over 1.00 m2	0.96	14.56	0.64	**15.20**	m2
101	Windows in panes:					
	small - not exceeding 0.10 m2	1.64	25.04	0.84	**25.88**	m2
	medium - 0.10 - 0.50 m2	1.14	17.42	0.74	**18.16**	m2
	large - 0.50 - 1.00 m2	1.07	16.25	0.64	**16.89**	m2

Unit Rates

		Man-Hours	Net Labour Price £	Net Mats Price £	Net Unit Price £	Unit
	extra large - over 1.00 m2	0.96	14.56	0.54	**15.10**	m2
102	Edges of opening casements	0.17	2.54	0.10	**2.64**	m
103	Structural members:					
	over 300 mm girth	0.76	11.51	1.98	**13.49**	m2
	not exceeding 150 mm girth	0.22	3.38	0.28	**3.66**	m
	150 - 300 mm girth	0.33	5.07	0.57	**5.64**	m
	Note 104 not used					
105	Each side of ornamental railings, gates and the like (grouped together) measured both sides overall regardless of voids:					
	over 300 mm girth	0.49	7.43	1.62	**9.05**	m2
106	Pipes and conduits, ducting, trunking and the like:					
	over 300 mm girth	0.67	10.16	2.12	**12.28**	m2
	not exceeding 150 mm girth	0.20	3.05	0.34	**3.39**	m
	150 - 300 mm girth	0.30	4.57	0.63	**5.20**	m
107	Eaves gutters:					
	over 300 mm girth	0.67	10.16	1.98	**12.14**	m2
	not exceeding 150 mm girth	0.20	3.05	0.28	**3.33**	m
	150 - 300 mm girth	0.30	4.57	0.57	**5.14**	m
108	Staircases:					
	over 300 mm girth	0.73	11.16	1.98	**13.14**	m2
	not exceeding 150 mm girth	0.22	3.38	0.28	**3.66**	m
	150 - 300 mm girth	0.32	4.90	0.57	**5.47**	m

WOODWORK - EXTERNALLY

109	**One undercoat alkyd based paint, one coat alkyd gloss to primed wood surfaces**					
110	Signwriting, per 100 mm in height, masking and setting out:					
	Arial font	1.04	15.84	0.56	**16.40**	nr
	manifestation or plain logo	2.08	31.68	0.56	**32.24**	nr
111	**Prepare, knotting and stopping; one coat wood primer, one undercoat alkyd based paint, one coat alkyd based paint gloss finish.** Primer @ £7.13 ; Undercoat @ £5.97 ; Gloss @ £5.97 (All per 1 ltr)					
112	General surfaces:					
	over 300 mm girth	0.61	9.31	2.08	**11.39**	m2
	not exceeding 150 mm girth	0.18	2.71	0.29	**3.00**	m
	150 - 300 mm girth	0.27	4.07	0.58	**4.65**	m
	isolated; not exceeding 0.50 m2	0.36	5.48	1.00	**6.48**	Nr
113	Glazed doors and screens in panes:					
	small - not exceeding 0.10 m2	1.58	24.03	1.29	**25.32**	m2

Unit Rates

		Man-Hours	Net Labour Price £	Net Mats Price £	Net Unit Price £	Unit
	medium - 0.10 - 0.50 m2	1.07	16.25	1.06	**17.31**	m2
	large - 0.50 - 1.00 m2	0.96	14.56	0.86	**15.42**	m2
	extra large - over 1.00 m2	0.88	13.37	0.64	**14.01**	m2
114	Windows in panes:					
	small - not exceeding 0.10 m2	1.64	25.04	1.64	**26.68**	m2
	medium - 0.10 - 0.50 m2	1.14	17.42	1.51	**18.93**	m2
	large - 0.50 - 1.00 m2	1.07	16.25	1.29	**17.54**	m2
	extra large - over 1.00 m2	0.96	14.56	1.06	**15.62**	m2
115	Edges of opening casements	0.17	2.54	0.16	**2.70**	m
116	Frames, linings and associated mouldings:					
	over 300 mm girth	0.61	9.31	2.08	**11.39**	m2
	not exceeding 150 mm girth	0.18	2.71	0.29	**3.00**	m
	150 - 300 mm girth	0.27	4.07	0.58	**4.65**	m
	VARNISH ON WOODWORK - EXTERNALLY					
117	**Prepare; one coat preservative basecoat, two coats external grade varnish, wood surfaces, first coat to untreated surfaces** (*Preservative basecoat @ £11.99 per 1 ltr; External grade varnish @ £16.27 per 1 ltr*)					
118	General surfaces:					
	over 300 mm girth	0.56	8.57	3.41	**11.98**	m2
	not exceeding 150 mm girth	0.17	2.56	0.59	**3.15**	m
	150 - 300 mm girth	0.25	3.82	1.11	**4.93**	m
	isolated; not exceeding 0.50 m2	0.33	5.03	1.70	**6.73**	Nr
119	Glazed doors and screens in panes:					
	small - not exceeding 0.10 m2	1.50	22.86	2.07	**24.93**	m2
	medium - 0.10 - 0.50 m2	1.04	15.90	1.77	**17.67**	m2
	large - 0.50 - 1.00 m2	0.92	13.97	1.40	**15.37**	m2
	extra large - over 1.00 m2	0.84	12.76	1.10	**13.86**	m2
120	Windows in panes:					
	small - not exceeding 0.10 m2	1.55	23.62	2.74	**26.36**	m2
	medium - 0.10 - 0.50 m2	1.14	17.42	2.43	**19.85**	m2
	large - 0.50 - 1.00 m2	0.97	14.73	2.07	**16.80**	m2
	extra large - over 1.00 m2	0.88	13.37	1.77	**15.14**	m2
121	Edges of opening casements	0.15	2.30	0.14	**2.44**	m
122	Frames, linings and associated mouldings:					
	over 300 mm girth	0.56	8.57	3.67	**12.24**	m2
	not exceeding 150 mm girth	0.17	2.59	0.63	**3.22**	m
	150 - 300 mm girth	0.25	3.81	1.15	**4.96**	m

	Unit Rates	Man-Hours	Net Labour Price £	Net Mats Price £	Net Unit Price £	Unit
123	**Prepare; two coats Breather Paint, wood surfaces** (*Breather paint @ £11.06 per 1 ltr*)					
124	General surfaces:					
	over 300 mm girth	0.28	4.26	2.81	**7.07**	m2
	not exceeding 150 mm girth	0.07	0.99	0.44	**1.43**	m
	150 - 300 mm girth	0.12	1.83	0.66	**2.49**	m
	isolated; not exceeding 0.50 m2	0.14	2.13	1.22	**3.35**	Nr
125	Windows in panes:					
	small - not exceeding 0.10 m2	0.60	9.14	2.15	**11.29**	m2
	medium - 0.10 - 0.50 m2	0.45	6.85	1.70	**8.55**	m2
	large - 0.50 - 1.00 m2	0.42	6.40	1.43	**7.83**	m2
	extra large - over 1.00 m2	0.39	5.94	1.15	**7.09**	m2
126	Frames, linings and associated mouldings:					
	over 300 mm girth	0.26	3.96	2.81	**6.77**	m2
	not exceeding 150 mm girth	0.08	1.22	0.44	**1.66**	m
	150 - 300 mm girth	0.13	1.98	0.66	**2.64**	m
127	**Prepare; two coats semi-gloss wood stain** (*Sadolin 'Extra' @ £10.67 per 1 ltr*)					
128	General surfaces:					
	over 300 mm girth	0.26	3.96	1.32	**5.28**	m2
	not exceeding 150 mm girth	0.06	0.91	0.21	**1.12**	m
	150 - 300 mm girth	0.11	1.68	0.43	**2.11**	m
	isolated; not exceeding 0.50 m2	0.14	2.13	0.64	**2.77**	Nr
129	Glazed doors and screens in panes:					
	small - not exceeding 0.10 m2	0.55	8.38	1.00	**9.38**	m2
	medium - 0.10 - 0.50 m2	0.41	6.24	0.79	**7.03**	m2
	large - 0.50 - 1.00 m2	0.38	5.79	0.68	**6.47**	m2
	extra large - over 1.00 m2	0.36	5.48	0.58	**6.06**	m2
130	Windows in panes:					
	small - not exceeding 0.10 m2	0.60	9.14	1.00	**10.14**	m2
	medium - 0.10 - 0.50 m2	0.45	6.85	0.79	**7.64**	m2
	large - 0.50 - 1.00 m2	0.42	6.40	0.68	**7.08**	m2
	extra large - over 1.00 m2	0.39	5.94	0.58	**6.52**	m2
131	Frames, linings and associated mouldings:					
	over 300 mm girth	0.26	3.96	1.43	**5.39**	m2
	not exceeding 150 mm girth	0.07	1.07	0.21	**1.28**	m
	150 - 300 mm girth	0.12	1.83	0.32	**2.15**	m
132	**Prepare; three coats hardwood finish, wood surfaces** (*Hardwood finish @ £10.33 per 1 ltr*)					
133	General surfaces:					
	over 300 mm girth	0.45	6.85	1.79	**8.64**	m2
	not exceeding 150 mm girth	0.11	1.68	0.24	**1.92**	m
	150 - 300 mm girth	0.18	2.74	0.48	**3.22**	m

Unit Rates

		Man-Hours	Net Labour Price £	Net Mats Price £	Net Unit Price £	Unit
	isolated; not exceeding 0.50 m2	0.29	4.42	0.89	**5.31**	Nr
134	Glazed doors and screens in panes:					
	small - not exceeding 0.10 m2	1.25	19.04	0.38	**19.42**	m2
	medium - 0.10 - 0.50 m2	0.95	14.47	0.31	**14.78**	m2
	large - 0.50 - 1.00 m2	0.80	12.18	0.24	**12.42**	m2
	extra large - over 1.00 m2	0.68	10.36	0.18	**10.54**	m2
135	Windows in panes:					
	small - not exceeding 0.10 m2	1.30	19.80	0.51	**20.31**	m2
	medium - 0.10 - 0.50 m2	1.04	15.84	0.40	**16.24**	m2
	large - 0.50 - 1.00 m2	0.86	13.10	0.31	**13.41**	m2
	extra large - over 1.00 m2	0.72	10.97	0.23	**11.20**	m2
136	Frames, linings and associated mouldings:					
	over 300 mm girth	0.45	6.85	1.84	**8.69**	m2
	not exceeding 150 mm girth	0.11	1.68	0.25	**1.93**	m
	150 - 300 mm girth	0.18	2.74	0.50	**3.24**	m
	Note 137 to 199 not used					
	PAPERHANGING AND DECORATION					
200	**Prepare plastered surfaces, size, hang woodchip paper, P.C. £1.98 per roll; one priming coat and two full coats vinyl silk emulsion paint** (@ £5.44 (white) per 1 litre)					
201	Walls:					
	over 300 mm wide	0.74	11.27	2.31	**13.58**	m2
	3.50 - 5.00 m high where ceiling is of dissimilar finish; over 300 mm wide	0.82	12.47	2.31	**14.78**	m2
	in staircase areas over 300 mm wide	0.81	12.37	2.31	**14.68**	m2
202	Ceilings:					
	over 300 mm wide	0.82	12.47	2.31	**14.78**	m2
	3.50 - 5.00 m high over 300 mm wide	0.90	13.69	2.31	**16.00**	m2
	in staircase areas over 300 mm wide	0.89	13.54	2.31	**15.85**	m2
	PAVINGS EXTERNALLY					
203	One sealer coat, one full coat chlorinated rubber line marking paint:					
	over 300 mm girth	0.69	10.51	5.96	**16.47**	m2
	not exceeding 150 mm girth	0.14	2.10	0.90	**3.00**	m
	150 - 300 mm girth	0.28	4.20	0.95	**5.15**	m
	manifestation or numeral 600 mm high	1.95	29.70	3.97	**33.67**	Nr

TRADE AND GOVERNMENT ORGANISATIONS

BCIS 50 years celebrating excellence

CONSTRUCTION
BCIS PRICE DATA 2012

Comprehensive Building Price Book 2012
Major and Minor Works dataset

The Major Works dataset focuses predominantly on large 'new build' projects reflecting the economies of scale found in these forms of construction. The Minor Works Estimating Dataset focuses on small to medium sized 'new build' projects reflecting factors such as increase in costs brought about the reduced output, less discounts, increased carriage etc.

Item code: 18770
Price: £165.99

SMM7 Estimating Price Book 2012

This dataset concentrates predominantly on large 'new build' projects reflecting the economies of scale found in these forms of construction. The dataset is presented in SMM7 grouping and order in accordance with the Common Arrangement of Work Sections. New glazing section and manhole build-ups.

Item code: 18771
Price: £139.99

Alterations and Refurbishment Price Book 2012

This dataset focuses on small to medium sized projects, generally working within an existing building and reflecting the increase in costs brought about by a variety of factors including reduction in output, smaller discounts, increased carriage, increased supervision etc.

Item code: 18772
Price £109.99

Guide to Estimating for Small Works 2012

This is a unique dataset which shows the true power of resource based estimating. A set of composite built-up measured items are used to build up priced estimates for a large number of common specification extensions.

Item code: 19064
Price: £59.99

For more information call **+44 (0)870 333 1600** email **contact@bcis.co.uk** or visit
www.bcis.co.uk/bcispricebooks

BCIS is the Building Cost Information Service of **RICS** — the mark of property professionalism worldwide

TRADE & GOVERNMENT ORGANISATIONS

Asbestos Removal Contractors Association (ARCA)
237 Branston Road
Burton Upon Trent
Staffordshire
DE14 3BT

☎: 01283 531126
📠: 01283 568228
✉: info@arca.org.uk
🌐: www.arca.org.uk

British Board of Agrément (BBA)
Bucknalls Lane
Garston
Herts
WD25 9BA

☎: 01923 665300
📠: 01923 665301
✉: contact@bba.star.co.uk
🌐: www.bbacerts.co.uk

British Coatings Federation (BCF)
The Stables
Thorncroft Manor
Thorncroft Drive
Leatherhead, Surrey
KT22 8JB

☎: 01372 700848
📠: 01372 700851
✉: enquiry@bcf.co.uk
🌐: www.coatings.org.uk

British Pest Control Association
1 Gleneagles House
South Street
Vernon Gate
Derby
DE1 1UP

☎: 01332 294288
✉: enquiry@bpca.org.uk
🌐: www.bpca.org.uk

Building and Allied Trades Joint Industrial Council
Gordon Fisher House
14 - 15 Great James Street
London
WC1N 3DP

☎: 020 7242 7583
📠: 020 7404 0296
✉: central@fmb.org.uk
🌐: www.fmb.org.uk

TRADE & GOVERNMENT ORGANISATIONS

Building Cost Information Service (BCIS)
Parliament Square
London
SW1P 3AD

☎: 0870 333 1600
📠: 020 7334 3851
✉: contact@bcis.co.uk
🌐: www.bcis.co.uk

Building Research Establishment (BRE)
Bucknalls Lane
Garston
Watford
WD25 9XX

☎: 01923 664000
✉: enquiries@bre.co.uk
🌐: www.bre.co.uk

Building Research Establishment : Scotland (BRE)
Orion House, Scottish Enterprise Technology Park
East Kilbride
Glasgow
G75 0RD

☎: 01355 576200
✉: eastkilbride@bre.co.uk
🌐: www.bre.co.uk

Chartered Institute of Arbitrators (CIArb)
12 Bloomsbury Square
London
WC1A 2LP

☎: 020 7421 7444
📠: 020 7404 4023
✉: info@ciarb.org
🌐: www.ciarb.org

Construction Industry Research & Information Association (CIRIA)
Classic House
174-180 Old Street
London
EC1V 9BP

☎: 020 7549 3300
📠: 020 7253 0523
✉: enquiries@ciria.org
🌐: www.ciria.org

Construction Industry Training Board (CITB)
Head Office
Bircham Newton
Kings Lynn
Norfolk
PE31 6RH

☎: 0344 994 4455
📠: 0300 456 7587
✉: levy.grant@cskills.org
🌐: www.cskills.org

TRADE & GOVERNMENT ORGANISATIONS

Construction Products Association
The Building Centre
26 Store Street
London
WC1E 7BT

☎: 020 7323 3770
🖷: 020 7323 0307
✉: enquiries@constructionproducts.org.uk
🌐: www. constructionproducts.org.uk

Federation of Master Builders (FMB)
Gordon Fisher House
14-15 James Street
London
WC1N 3DP

☎: 020 7242 7583
🖷: 020 7404 0296
✉: central@fmb.org.uk
🌐: www.fmb.org.uk

Mineral Products Association (MPA)
Gillingham House
38-44 Gillingham Street
London
SW1V 1HU

☎: 020 7963 8000
🖷: 020 7963 8001
✉: info@mineralproducts.org
🌐: www.mineralproducts.org

National Assess and Scaffolding Confederation (NASC)
4th Floor
12 Bridewell Place
London
EC4V 6AP

☎: 020 7822 7400
🖷: 020 7822 7401
✉: enquiries@nasc.org.uk
🌐: www.nasc.org.uk

National Federation of Builders Limited (NFB)
B & CE Building
Manor Royal
Crawley
West Sussex
RH10 9QP

☎: 08450 578160
🖷: 08450 578161
✉: national@builders.org.uk
🌐: www.builders.org.uk

TRADE & GOVERNMENT ORGANISATIONS

National Statistics Library and Information Service
Government Buildings
Cardiff Road
Newport
NP10 8XG

☎:0845 601 3034
📠: 01633 652747
✉: info@statistics.gov.uk
🌐: www.statistics.gov.uk

Painting and Decorating Association
32 Coton Road
Nuneaton
Warwickshire
CV11 5TW

☎: 024 7635 3776
📠: 024 7635 4513
✉: info@paintingdecoratingassociation.co.uk
🌐: www.paintingdecoratingassociation.co.uk

PRA Coatings Technology Centre
14 Castle Mews
High Street, Hampton
Middlesex
TW12 2NP

☎: 020 8487 0800
📠: 020 8487 0801
✉: coatings@pra.org.uk
W: www.pra-world.com

Property Care Association (PCA)
Lakeview Court
Ermine Business Park
Huntington, Cambs
PE29 6XR

☎: 0844 375 4301
📠: 01480 417 587
✉: pca@property-care.org
🌐: www.property-care.org

Royal Incorporation of Architects in Scotland (RIAS)
15 Rutland Square
Edinburgh
EH1 2BE

☎: 0131 229 7545
📠: 0131 228 2188
✉: info@rias.org.uk
🌐: www.rias.org.uk

Royal Institute of British Architects (RIBA Enterprises)
66 Portland Place
London
W1B 1AD

☎: 020 7580 5533
📠: 020 7255 1541
✉:info@riba.org
🌐:www.architecture.com

TRADE & GOVERNMENT ORGANISATIONS

Royal Institution of Chartered Surveyors (RICS)
Parliament Square
London
SW1P 3AD

☎: 0870 333 1600
📠: 020 7 334 3811
✉: contactrics@rics.org
🌐: www.rics.org

Also At:
Surveyor Court
Westwood Way
Coventry
CV4 8JE

☎: 0870 333 1600
📠: 0207 334 3811
✉: contactrics@rics.org
🌐: www.rics.org

9 Manor Place
Edinburgh
EH3 7DN
Scotland

☎: 0131 225 7078
📠: 0131 240 0830
✉: scotland@rics.org
🌐: www.rics.org/scotland

9 - 11 Corporation Square
Belfast
BT1 3AJ
Northern Ireland

☎: 028 9032 2877
📠: 028 9023 3465
✉: ricsni@rics.org
🌐: www.rics.org/northernireland

7 St Andrews Place
Cardiff
CF10 3BE
Wales

☎: 029 2022 4414
📠: 029 2022 4416
✉: wales@rics.org
🌐: www.rics.org/wales

Scottish Building Federation
Crichton House, 4 Crichton's Close
Holyrood, Edinburgh
EH8 8DT

☎: 0131 556 8866
📠: 0131 558 5247
✉: info@scottish-building.co.uk
🌐: www.scottish-building.co.uk

Scottish Decorators' Federation
Castlecraig Business Park
Players Road
Stirling
FK7 7SH

☎: 01786 448838
📠: 01786 450541
✉: info@scottishdecorators.co.uk
🌐: www.scottishdecorators.co.uk

TRADE & GOVERNMENT ORGANISATIONS

The Association of Building Engineers (ABE)
Lutyens House, Billing Brook Road
Weston Favell
Northampton, Northants
NN3 8NW

☎: 0845 126 1058
📠: 01604 784220
✉: building.engineers@abe.org.uk
🌐: www.abe.org.uk

The Building Centre (Exhibitions and Bookshop)
26 Store Street
London
WC1E 7BT

☎: 020 7692 4000
✉: reception@buildingcentre.co.uk
🌐: www.buildingcentre.co.uk

The Guild of Master Craftsmen
166 High Street
Lewes
East Sussex
BN7 1XU

☎: 01273 478449
📠: 01273 478606
✉: theguild@thegmcgroup.com
🌐: www.guildmc.com

Wood Protection Association
5C Flemming Court
Castleford
West Yorkshire
WF10 5HW

☎: 01977 558274
✉: info@wood-protection.org
🌐: www.wood-protection.org